Lecture Notes in Earth Sciences 132

Editors:

J. Reitner, Göttingen
M. H. Trauth, Potsdam
K. Stüwe, Graz
D. Yuen, USA

Founding Editors:

G. M. Friedman, Brooklyn and Troy
A. Seilacher, Tübingen and Yale

T0181208

For further volumes:
http://www.springer.com/series/772

Lecture Notes in Earth Sciences

132

Editors

Founding Editors

Saswati Bandyopadhyay
Editor

New Aspects of Mesozoic Biodiversity

 Springer

Editor
Saswati Bandyopadhyay
Geological Studies Unit
Indian Statistical Institute
203 B. T. Road
Kolkata 700 108
India
saswati@isical.ac.in
saswati2608@gmail.ac.in

ISBN 978-3-642-26390-3   ISBN 978-3-642-10311-7(eBook)
DOI 10.1007/978-3-642-10311-7
Springer Heidelberg Dordrecht London New York

Cover design: Integra Software Services Pvt. Ltd., Pondicherry

Printed on acid-free paper

Springer is part of Springer Science+Business Media (www.springer.com)

Foreword

The Indian Statistical Institute (ISI) was established on 17th December, 1931 by a great visionary Prof. Prasanta Chandra Mahalanobis to promote research in the theory and applications of statistics as a new scientific discipline in India. In 1959, Pandit Jawaharlal Nehru, the then Prime Minister of India introduced the ISI Act in the parliament and designated it as an Institution of National Importance because of its remarkable achievements in statistical work as well as its contribution to economic planning.

Today, the Indian Statistical Institute occupies a prestigious position in the academic firmament. It has been a haven for bright and talented academics working in a number of disciplines. Its research faculty has done India proud in the arenas of Statistics, Mathematics, Economics, Computer Science, among others. Over seventy five years, it has grown into a massive banyan tree, like the institute emblem. The Institute now serves the nation as a unified and monolithic organization from different places, namely Kolkata, the Headquarters, Delhi, Bangalore, and Chennai, three centers, a network of five SQC-OR Units located at Mumbai, Pune, Baroda, Hyderabad and Coimbatore, and a branch (field station) at Giridih.

The platinum jubilee celebrations of ISI have been launched by Honorable Prime Minister Prof. Manmohan Singh on December 24, 2006, and the Govt. of India has declared 29th June as the "Statistics Day" to commemorate the birthday of Prof. Mahalanobis nationally.

Prof. Mahalanobis, was a great believer in interdisciplinary research, because he thought that this will promote the development of not only Statistics, but also the other natural and social sciences. To promote interdisciplinary research, major strides were made in the areas of computer science, statistical quality control, economics, biological and social sciences, physical and earth sciences.

The Institute's motto of "unity in diversity" has been the guiding principle of all its activities since its inception. It highlights the unifying role of statistics in relation to various scientific activities.

In tune with this hallowed tradition, a comprehensive academic programme, involving Nobel Laureates, Fellows of the Royal Society, Abel prize winner and other dignitaries, has been implemented throughout the Platinum Jubilee year, highlighting the emerging areas of ongoing frontline research in its various scientific divisions, centers, and outlying units. It includes international and national-level seminars, symposia, conferences and workshops, as well as series of special lectures. As an outcome of these events, the Institute is bringing out a series of comprehensive volumes in different subjects including those published under the title *Statistical Science and Interdisciplinary Research* of the World Scientific Press, Singapore.

The present volume titled "New Aspects of Mesozoic Biodiversity" is one such outcome published by Springer-Verlag. It deals with certain aspects of terrestrial biodiversity during the Mesozoic era. This specialized volume has seven chapters written by eminent palaeontologists from some of the pioneering groups of the world. The contributions broadly range from contemporary issues of fossil vertebrates including new insights of tetrapod evolution during Palaeozoic and Mesozoic eras, discovery of new vertebrate fossils from different continents, new information on palaeo-osteohistology, to palaeobiogeography of the Indian plate during the end Mesozoic. I believe the state-of-the art studies presented in this book will be very useful to researchers as well as practitioners.

Thanks to the contributors for their excellent research contributions, and to the volume editor Dr. Saswati Bandyopadhyay for her sincere effort in bringing out the volume nicely. Thanks are also due to Springer-Verlag for their initiative in publishing the book and being a part of the Platinum Jubilee endeavor of the Institute.

Kolkata
December 2009

Sankar K. Pal
Director
Indian Statistical Institute

Preface

This book encompasses some aspects of terrestrial biodiversity during the Mesozoic era. Biodiversity is a rapidly growing field of study which, more comprehensively, involves the evolution of the life in an evolving earth, and these two cannot be decoupled, though many of us tend to focus on one side or the other of this fascinating story. Terrestrial ecosystem ushered in during the Palaeozoic with the Cambrian explosion giving rise to the major invertebrate phyla dominating the Palaeozoic scenario followed up by the appearance of animals with backbones during the Late Palaeozoic. The Early Palaeozoic had elevated atmospheric CO_2 and warm temperatures extended to high latitudes but due to the extensive continental glaciations Gondwanaland became ice-covered during the later part of the Palaeozoic. However, warm climate prevailed in large areas of Laurasia. Towards the end of the Palaeozoic the cold Gondwanaland became warmer and the vegetation flourished. In Mesozoic the climate was generally dry and warm with seasonal monsoon in low latitude regions and the scenario of terrestrial ecosystem during this time had considerably changed. Though Mesozoic started with an impoverished diversity level due to the great end-Permian mass extinction event, it witnessed the emergence of a variety of terrestrial vertebrates which flourished, proliferated and took over the set scenario in the nearly vacant ecological niches and gradually increased in diversity reaching a new peak towards the end of the Mesozoic. This era actually shaped the founding of the present terrestrial ecosystem.

During the last two decades new discoveries and new techniques have led to better understanding of the diverse terrestrial faunas of the Mesozoic and their evolution in the backdrop of changing palaeogeography. The present book is the outcome of the "International Conference on Geology: Indian Scenario and Global Context" held at the Indian Statistical Institute as part of their Platinum Jubilee celebration in 2008 which coincided with the Golden Jubilee of the Geological Studies Unit of I. S. I.. The session on the "Evolution of Late Palaeozoic and Mesozoic Terrestrial Vertebrates" was organized on January 9, 2008; a good number of geologists and palaeobiologists from different parts of the world including several students attended the conference.

In a series of seven papers this book reflects the broader perspective of evolution of life and land during the Mesozoic with a global view. It contributes papers on contemporary issues of fossil vertebrates including new insights of tetrapod evolution during Palaeozoic and Mesozoic eras, discovery of new vertebrate fossils from different continents, new information on palaeo-osteohistology and palaeobiogeography of the Indian plate during the end Mesozoic.

In the first chapter, Kemp has presented the current major advances in the palaeobiological researches on vertebrates of Late Palaeozoic and Mesozoic time; in his review; he focused on (i) new discoveries such as feathered basal birds from China, snake with a well-defined sacrum supporting a pelvis and functional hind limbs outside its ribcage from Argentina etc. from different corners of the world, filling the gaps in the evolution of higher tetrapod taxa and supplementing information in the configuration of faunal turnover and palaeo-community structure, (ii) application of new techniques including molecular palaeontology, CT scanning of fossil bones, stable isotope analysis and (iii) the developments of new concepts such as correlated progression. The reptilian group, Lepidosauria with its two major components, Rhynchocephalia and Squamata, has a long evolutionary history since Triassic till today though patchy at times; Evans and her co-author made an analysis of the evolution, diversification and extinction of these two groups in Gondwanaland and Laurasia through ages.

Novas and his co-authors described a new abelisaurid dinosaur from the Upper Cretaceous Lameta Formaton of India and discussed its affinity to other Gondwanan abelisauroids. Recovery of a new primitive pterosaur from the continental Late Triassic Caturita Formation of Brazil has been communicated by Bonaparte and his co-authors who suggested that the early pterosaur evolved both in the terrestrial and littoral marine environment. Ray and her co-authors examined the bone microstructure of a Triassic kannemeyeriid dicynodont from India and showed that there are three distinct ontogenic stages of this taxon.

The last two papers deal with the Indian fauna, their evolution, radiation, dispersal and extinction during the end Mesozoic when the northward drifting Indian plate became isolated from the rest of Gondwanaland. Sahni has summarized all the Late Mesozoic biota from India, their endemism, cosmopolitanism and their implications for evolution in the background of end Cretaceous mass extinction event as well as the extensive Deccan basalt eruption. Chatterjee and his co-author have put together the information on the tectonic evolution of the Indian plate with its biogeography during the Cretaceous-Tertiary period testing the models of geodispersal and vicariance.

All the manuscripts were peer-reviewed by two or more experts whose comments and suggestions were carefully attended. I would like to thank S. Apesteguia (MACN, Buenos Aires, Argentina), P. M. Barrett (NHM, London, UK), T. Bhattacharya (University of Calcutta, India), S. Burch (Stony Brook University, USA), M. T. Carrano (Smithsonian Institution, Washington, DC, USA), A. Chinsamy-Turan (University of Cape Town, South Africa), D. D. Gillette (MNA, Flagstaff, USA), J. D. Harris (Dixie State College, Utah, USA), T. S. Kemp (University of Oxford, UK), D. W. Krause (Stony Brook University, USA), D. Norman (University of Cambridge, UK), K. Padian (University of California, Berkeley), G. V. R. Prasad (University of Delhi, India), V. H. Reynoso (Instituto de Biologia UNAM, Mexico), A. Sahni (Panjab University, Chandigarh, India) and

P. Upchurch (University College of London, UK) for reviewing the manuscripts and for their constructive suggestions. I am grateful to the authors for their cooperation and patience with editing process. I would like to express my gratitude to C. Bendall and J. Sterritt-Brunner of Springer-Verlag for their help and co-operation. Thanks are due to P. S. Ghosh, S. N. Sarkar, T. RoyChowdhury and D. P. Sengupta of Indian Statistical Institute for their constant help and encouragements.

Kolkata, India   Saswati Bandyopadhyay

Contents

Contributors

Ravi Appana Department of Geology and Geophysics, Indian Institute of Technology, Kharagpur 721302, India, appan002@umn.edu

Saswati Bandyopadhyay Geological Studies Unit, Indian Statistical Institute, Kolkata 700108, India, saswati@isical.ac.in

J. F. Bonaparte Division Paleontología, Fundación de Historia Natural "Félix de Azara", Buenos Aires 1405, Argentina, bonajf@speedy.com.ar

Sankar Chatterjee Department of Geosciences, Museum of Texas Tech University, Lubbock, TX 79409, USA, sankar.chatterjee@ttu.edu

P. M. Datta Palaeontology Division, Geological Survey of India, Eastern Region, Kolkata 700091, India, shiladutta59@yahoo.in

Susan E. Evans Research Department of Cell and Developmental Biology, University College London, London WC1E 6BT, UK, ucgasue@ucl.ac.uk

Marc E. H. Jones Research Department of Cell and Developmental Biology, University College London, London WC1E 6BT, UK, marc.jones@ucl.ac.uk

T. S. Kemp St. John's College and Oxford University Museum of Natural History, Oxford OX1 3PW, UK, tom.kemp@sjc.ox.ac.uk

Fernando E. Novas Conicet, Museo Argentino de Ciencias Naturales, Buenos Aires 1405, Argentina, fernovas@yahoo.com.ar

Sanghamitra Ray Department of Geology and Geophysics, Indian Institute of Technology, Kharagpur 721302, India, sray@gg.iitkgp.ernet.in

Dhiraj K. Rudra Geological Studies Unit, Indian Statistical Institute, Kolkata 700108, India, dhirajr75@gmail.com

A. Sahni Centre of Advanced Study in Geology, Panjab University, Chandigarh 160014, India, ashok.sahni@gmail.com

C. L. Schultz IGEO, UFRGS, Porto Alegre, RS 91501-970, Brazil, cesar.schultz@ufrgs.br

Christopher Scotese Department of Geology, University of Texas, Arlington, TX 76019-0049, USA, cscotese@uta.edu

M. B. Soares IGEO, UFRGS, Porto Alegre, RS 91501-970, Brazil, marina.soares@ufrgs.br

Chapter 1
New Perspectives on the Evolution of Late Palaeozoic and Mesozoic Terrestrial Tetrapods

T.S. Kemp

1.1 Introduction

Palaeobiology contributes in principle to evolutionary theory by providing evidence about several phenomena of long-term evolution that are only revealed at all by the fossil record. Any theory about the mechanism of evolution that claims to be comprehensive must be able to account for such geological-timescale events, including: the several million years of stasis that is typical of palaeospecies; the great kaleidoscopic pattern of taxonomic turnover at every level from species and genera to orders and classes; the occasional periods of mass extinction during which anything up to 90% of the Earth's species disappear; and the environmental circumstances surrounding the major morphological transitions represented by the appearance of new higher taxa. The fundamental issue is whether simple extrapolation of mechanisms known to occur at the level of the interbreeding population – Darwinian natural selection in particular – provides a sufficient explanation for the long term course of phylogenetic change, or whether rare events or extremely slow processes that are unobservable in field or laboratory are also at work over this timescale. There has been a regrettable failure even to take seriously this possibility on the part of some authors (e.g. Charlesworth, 1996; Bell, 2000); others however are well aware of the issue (e.g. Gould, 1994; Kemp, 1999; Grantham, 2007).

The fossil record of Late Palaeozoic and Mesozoic terrestrial tetrapods offers some of the most important evidence of all concerning aspects of long term evolutionary patterns. During this time, several major evolutionary transitions occurred that resulted in new higher taxa, and for which several intermediate grades between the ancestral and the descendant are represented as known fossils. Following the origin of the taxon Tetrapoda itself during the Late Devonian, there duly appeared the lineages leading to such radically new kinds of tetrapods as the amphibians, turtles, mammals, snakes, dinosaurs, and birds. This period of time is also uniquely important for palaeobiogeographical study, because great changes in the continental

T.S. Kemp (✉)
St. John's College and Oxford University Museum of Natural History, Oxford OX1 3PW, UK
e-mail: tom.kemp@sjc.ox.ac.uk

S. Bandyopadhyay (ed.), *New Aspects of Mesozoic Biodiversity*,
Lecture Notes in Earth Sciences 132, DOI 10.1007/978-3-642-10311-7_1,
© Springer-Verlag Berlin Heidelberg 2010

configurations coincided with the diversification of several of these new tetrapod taxa, which throws light upon the relationship between patterns of phylogeny and patterns of biogeographic vicariance (e.g. Upchurch et al., 2002).

Like all sciences, progress in palaeobiology may be described as occurring on three distinguishable though intimately interrelated fronts. First there is the accumulation of new, empirically derived information resulting from the discovery of new fossils, and new data about their palaeobiogeographical distribution and palaeoecological setting. Second there are new techniques for studying existing material that are capable of generating more detailed, accurate answers to palaeobiological questions than hitherto possible. Third there is the development of new concepts or frameworks for thinking about the fossil record, leading to the generation of novel hypotheses and theories about evolution at this scale.

1.2 New Fossils

Without doubt, the most spectacular new fossils of the Mesozoic are the tetrapods of the Yixian Formation of China (Zhang, 2006). This remarkable locality is dated as Early Cretaceous, probably Hauterivan to Aptian, and according to Zhou et al. (2003), the palaeofauna resulted from a combination of a shallow lacustrine environment and rapid volcanic ashfall. A series of small tetrapods are superbly preserved, often including impressions of pelt or feathers. There are small nonavian theropod dinosaurs, including feathered specimens, and basal birds. Complete mammalian skeletons of several taxa are found including, importantly, the early metatherian *Sinodelphys* (Luo et al., 2003) and early eutherian *Eomaia* (Ji et al., 2002). These two fossils helpfully place the latest divergence date of the marsupials from the placentals at about 120 million years ago.

However aesthetically pleasing these specimens are, from the perspective of evolutionary theory, the most important recent finds of Late Palaeozoic and Mesozoic tetrapods are firstly those that are stem members of lineages leading to major new taxa, and secondly those that significantly expand the known diversity, morphological disparity, and temporo-spatial distribution of ecologically important extinct taxa. Among the many discoveries of the last few years bearing on these questions, the following are particularly illustrative.

1.2.1 Sequence of Acquisition of Characters: Stem Tetrapods

Until the early 1990s, almost everything known from the fossil record about the transition from ancestral fish-grade tetrapodamorph to fully limbed tetrapod was based on the comparison of the Late Devonian *Eusthenopteron* as an "ancestral fish" with *Ichthyostega* as a basal tetrapod (Save-Soderbergh, 1932; Jarvik, 1980), although a certain amount was also known about the more progressive "fish" grade form *Panderichthys* (Vorobyeva and Schultze, 1991; Boisvert, 2005). The next important addition came from the detailed description of *Acanthostega* (Clack, 1994; Coates, 1996), which is more basal than *Ichthyostega*. Meanwhile, several other Late

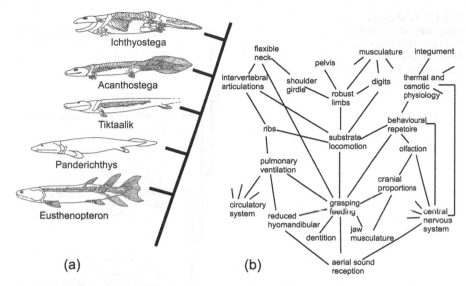

Fig. 1.1 (**a**) Cladogram of the best known stem tetrapods. (**b**) Functional and structural integration between the parts of an evolving tetrapod, to show the principle of the correlated progression model of the origin of major new taxa. From Kemp (2007a)

Devonian tetrapod or near tetrapod genera have been described from much more fragmentary material, such as *Ventastega* from Latvia, *Tulerpeton* from Russia, and *Elginerpeton* from Scotland, as reviewed by Clack (2002).

Cladistic analysis of these various forms gave a good deal of information about the sequence of acquisition of tetrapod characters within the hypothetical lineage of ancestors and descendants, but there remained a substantial morphological gap between what were still essentially finned "fish" and digit-bearing tetrapods. Therefore the recent description of *Tiktaalik* (Daeschler et al., 2006; Shubin et al., 2006), which partially spans this gap, adds another highly informative stage in the sequence, refining yet further what can be inferred about the morphological evolution of tetrapods (Fig. 1.1a). Indeed, *Tiktaalik* has been compared with *Archaeopteryx* in the importance of its particular combination of ancestral and derived characters, and therefore in its role of further resolving the sequence of acquisition of tetrapod traits. Any comprehensive account of the origin of tetrapods and their transition from aquatic to terrestrial habitat must necessarily start with this information.

1.2.2 Sequence of Acquisition of Characters: Stem Dicynodontian Therapsids

The dicynodontian therapsids of the Late Permian were of enormous evolutionary significance because they were the first highly abundant terrestrial, herbivorous, often herd-dwelling tetrapods, a mode of life that was to be pursued successively in the Mesozoic by the ornithischian dinosaurs, and in the Tertiary by the ungulate

Fig. 1.2 Cladogram of three basal anomodonts and the dicynodontian *Eodicynodon*. Redrawn from Reisz and Sues (2000), Modesto et al. (1999), and Rubidge (1990)

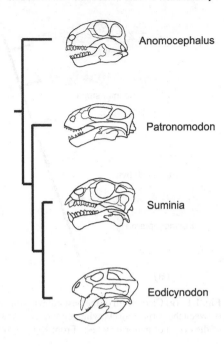

placentals and diprotodont marsupials. Like these taxa, dicynodontians were also highly diverse, with numerous species showing relatively minor differences in feeding and locomotory structures from one another. Morphologically, dicynodontians were highly modified from the basal therapsid form, and a series of recently described mid-Permian fossils from the South African Karoo and Russia have gone far towards illustrating the manner in which they achieved their specialisation (Fig. 1.2). The most basal of these is *Anomocephalus* (Modesto et al., 1999), which is relatively long-snouted and lacks the extensive re-modelling of the adductor musculature seen in more derived forms. *Patronomodon* (Rubidge and Hopson, 1996) has an enlarged temporal fenestra, and depressed jaw articulation region of the skull. *Suminia* (Rybczynski, 2000) has taken these trends further, with a dorsally bowed zygomatic arch and a jaw hinge allowing antero-posterior shifts of the lower jaw. *Eodicynodon* (Rubidge, 1990), which has been known for some time, is more or less fully dicynodontian in structure, having evolved the characteristic reorganisation of the jaw musculature and lost the anterior teeth apart from a pair of upper tusks.

The importance of this sequence is that it greatly increases the understanding of how, anatomically and functionally, the uniquely novel dicynodontian feeding structures evolved (Reisz and Sues, 2000; Kemp, 2005).

1.2.3 Sequence of Acquisition of Characters: Other Tetrapods

The anatomical and functional evolution of birds is beginning to be better understood as a consequence of a plethora of relevant discoveries of feathered dinosaurs, stem-birds and early avians, notably in the Early Cretaceous Jehol fauna (Zhou, 2004; Xu and Norrell, 2006; Turner et al., 2007; Hu et al., 2009).

Another major tetrapod taxon for which Mesozoic intermediate stages have recently come to light is the snakes. *Pachyrachis* (Caldwell and Lee, 1999), for example, has small but distinct hind limbs, while the more recently discovered *Najash* (Apesteguía and Zaher, 2006) has even larger ones, here associated with a pelvis still connected to the sacral vertebrae and apparently capable of a degree of locomotory function. Even more remarkable is *Odontochelys*, a Late Triassic turtle in which the carapace is represented only by expanded ribs and neural plates, evidently an intermediate stage towards the definitive chelonian structure (Li et al., 2008).

1.2.4 Morphological Disparity: New Kinds of Dinosaurs

One of the ways to throw light upon the nature, laws, and potential of morphological structure is to increase the range of known actual morphologies, and from time to time strange new fossil forms perform this purpose. No extinct taxon is more prone to this than the dinosaurs, and descriptions of new, highly aberrant species continue to appear regularly. The 3–4 m high giant, ostrich-like theropod *Gigantoraptor* (Xu et al., 2007), and *Nigersaurus*, a sauropod with an extraordinary, paper-thin skull and transversely oriented rows of fine teeth (Sereno et al., 2007) are just such surprises. Even more unpredictable, the four-winged, bird-like *Microraptor* from the Yixian has challenged theories on the origin of flight (Padian and Dial, 2005; Chatterjee and Templin, 2007).

1.2.5 Pattern of Ecological Replacement: The Rise of the Dinosaurs

The origin and early diversification of dinosaurs in the Upper Triassic is an area of perennial interest, given the Mesozoic dominance of the group from the Jurassic onwards. Most of the new work on dinosaurs has actually been re-description and exceedingly detailed phylogenetic analysis of existing material (e.g. Rauhut, 2003; Butler, 2005; Langer and Benton, 2006; Upchurch et al., 2007). Additional to this, however, other recent discoveries bear on the more general palaeobiological question concerning the process of replacement of the basal archosaurs and other terrestrial taxa such as rhynchosaurs and cynodonts by Dinosauria during the Late Triassic. It has long been debated whether this was a competitive process in which dinosaur species were in some way competitively superior, or an opportunistic one whereby dinosaurs only diversified after an environmental perturbation had caused the extinction of the other groups (Benton, 1996; Kemp, 1999). Testing between these two models is not easy, and depends on an estimate of the exact temporal relationship between the decline of the old and the increase of the new taxa, and whether the event was accompanied by palaeoecological signals of an ecological perturbation that could feasibly account for an initial extinction event. The evidence from the time course of the replacement has tended to support the opportunistic model, in which an extinction event at the close of the Carnian saw the end of the hitherto dominant non-dinosaur herbivore groups, while dinosaurs did not radiate extensively until the succeeding Norian (Benton, 1994). However, this interpretation is disturbed by the recent discovery of North American Upper Triassic representatives of several taxa of basal dinosauriforms that were previously known only from the Middle Triassic (Irmis et al., 2007). This evidence for a significant overlap between these basal groups and the dinosaurs indicates that a much longer, more gradual process was involved, suggesting that a form of competitiveness played a greater role. One attractive possibility is that it was a case of "incumbent replacement", lasting throughout the Upper Triassic, in which the rate of taxon replacement was controlled by the rate of background extinction of individual species of the pre-existing community, rather than by direct species to species competition (Rosenzweig and McCord, 1991). Other recently described fossil evidence from both North America (Lucas and Tanner, 2006) and India (Bandyopadhyay and Sengupta, 2006) indicates that the replacement was completed by about the close of the Norian.

1.2.6 Ecological Potential: The Disparity of Mesozoic Mammals

The mammals of the Jurassic and Cretaceous (Kielan-Jaworowska et al., 2004) have always been believed to be an ecologically conservative taxon of small, insectivorous and omnivorous animals, analogous in habits to the modern insectivores, rodents, and small opossums of the modern world, but vastly less diverse and abundant. However, in the last few years a most surprising range of adaptive types has been described. With an estimated body weight of 13 kg and a presacral length of 700 mm, *Repenomamus giganticus* (Hu et al., 2005) was considerably larger than was thought possible for Mesozoic mammals, and indeed shows evidence of having fed on young dinosaurs. *Fruitafossor* (Luo and Wible, 2005) is adapted for a fossorial existence with powerful, mole-like limbs. *Castorocauda* (Ji et al., 2006) pursued a beaver-like aquatic life, and *Volticotherium* (Meng et al., 2006) was a gliding mammal. It is thus becoming clear that most of the range of habitats occupied by modern small mammals also occurred in the archaic Mesozoic groups, and that during this Era small mammals had a far more significant ecological role in the terrestrial community than hitherto supposed (Jin et al., 2006).

1.3 New Techniques

1.3.1 The Molecular Revolution: Phylogenetic Reconstruction

No area of biology has been left untouched by the last decade's avalanche of data resulting from the development of routine sequencing of nucleic acids. For phylogenetic studies, the sheer amount of information contained in DNA, the objectivity of defining a unit character as a single nucleotide, and the availability of ever-more sophisticated statistical methods for analysing it lead to far more resolved and precisely dated phylogenetic branching points than morphology has proved capable of revealing. Of course molecular systematics applies directly only to recent organisms, but it can have an indirect effect on the phylogenetic analysis of the extinct members of modern taxa whose stem-groups and early divergencies are represented in the fossil record. Moreover, molecular techniques necessarily have an effect on the confidence that can be placed on phylogenies of taxa that are entirely fossils. In several cases, such as the interrelationships of placental mammals and modern birds, the molecular evidence has shown that morphology alone is incapable of recapturing the phylogeny. By analogy, this should greatly increase our scepticism about the reliability of the morphological-based phylogenies of certain entirely fossil groups where the morphological support is not great.

The rise to dominance of molecular over morphological data for phylogenetic reconstruction and its effect on interpretation of the Mesozoic tetrapod fossil record is illustrated most comprehensively by the case of the placental mammals. The earliest member of the Eutheria, which includes the stem-group plus the crown-group placentals, is *Eomaia* (Ji et al., 2002). It is probably Barremian in age, around 125 Ma, and occurs in the Yixian Formation of China, so is Laurasian in distribution. A considerable variety of taxa of eutherians are known from the Aptian-Albian and onwards into the Late Cretaceous, most abundantly in Asia and North America but also including a small number in Africa and India. However, no undisputed member of any of the modern placental orders (crown placentals) has been described prior to the Cretaceous-Tertiary boundary, and almost all make their appearance in the fossil record in a window of time between about 65 and 55 Ma, from Early Palaeocene to Early Eocene. Concerning the interrelationships between the placental orders, morphological analysis generated relatively few supraordinal groups (Fig. 1.3), and even the monophyletic status of these was always subject to dispute (e.g. Novacek et al., 1988; McKenna and Bell, 1997; Rose, 2006). Otherwise, the phylogeny was dominated by a large, unresolved polytomy of up to 10 lineages.

From the late 1990s onwards, rapidly accumulating molecular sequence data has resulted in a totally unpredicted, radical modification to the morphological-based view of placental interrelationships (Springer et al., 2003, 2005). Most of the morphological-based monophyletic groups have been either rejected outright, or had their membership altered, whilst the central polytomy has been fully resolved (Fig. 1.3). Beyond all reasonable doubt, the placental orders fall into the now familiar four superorders Afrotheria, Xenarthra, and the boreotherian sister groups Laurasiatheria and Euarchontoglires. Equally unpredicted, the estimated dates of most of the ordinal divergences based on the molecular evidence and using a variety of clock-like and relaxed clock-like models (Fig. 1.4), have been pushed back into the Cretaceous, some by a mere 10 Ma or so, others much farther (Springer et al., 2003, 2005; Bininda-Emonds et al., 2007).

The immediate response to the new phylogeny by a number of palaeobiologists was, predictably enough, that the molecular evidence must be incorrect, and

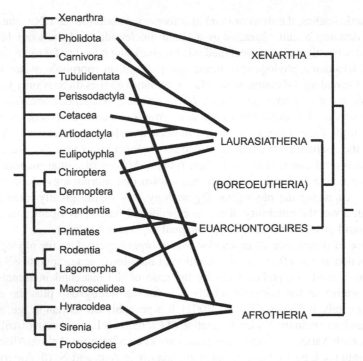

Fig. 1.3 A comparison of Novacek et al.'s (1988) morphological based interrelationships of placental mammalian orders (*left*), with Springer et al.'s (2003, 2005) molecular-based phylogeny (*right*), showing how radical a modification the molecular data caused

particularly as regards the new estimates of the divergence dates. However, given the volume of data now supporting it, the molecular-based phylogeny itself is by far the best supported hypothesis of relationships; mammalian palaeobiologists now have the exciting task of reinterpreting the anatomical evolution as inferred from the new cladogram, including seeking morphological characters that are congruent with the molecular-based groupings (Asher et al., 2003), and elucidating the historical biogeography of placental mammals (Archibald, 2003; Kemp, 2005; Hunter and Janis, 2006).

More problematic, and therefore more challenging is the matter of the divergence dates. If the molecular-based dates are anything like correct, then why are no placental orders represented prior to the end of the Cretaceous? But if the fossil-based divergence dates are more accurate, then the rate of molecular evolution must be variable to an as yet inexplicable extent. It is always possible that crown placentals were too rare to have been discovered as fossils, or that they were diversifying in a region of the world not represented by Late Cretaceous continental sediments, such as part of Gondwana, but there are good statistical arguments against these explanations (Foote et al., 1999; Donaghue and Benton, 2007). A more interesting possible explanation is the "Long Fuse" hypothesis that crown placentals were in fact present in the Late Cretaceous and are indeed represented by known fossils, but

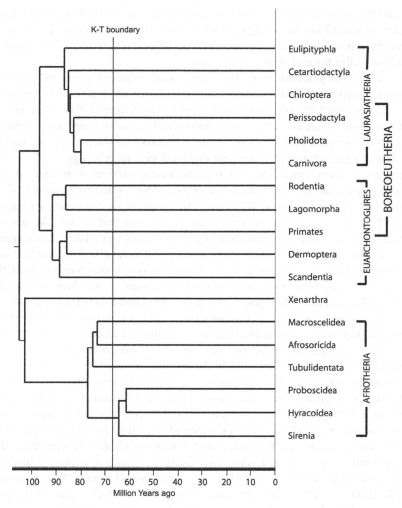

Fig. 1.4 Molecular-based estimates of dates of divergence of placental superorders and orders (redrawn from Murphy and Eizirik, 2009)

that so little morphological divergence had occurred that their affinities are difficult to recognise. At least three possible Cretaceous crown placentals have been suggested: zalambdalestids as members of Glires (rodents plus lagomorphs), zhelestids as assorted ungulate orders, and palaeoryctids as members of the Carnivora and Creodonta (Archibald et al., 2001). If the long-fuse hypothesis is true, then it raises the challenging question of why the original lineage splitting was separated from major morphological evolution. Perhaps rapid diversification at a low taxonomic level in the Late Cretaceous was associated with a new ecological opportunity to divide the small, nocturnal insectivore/omnivore habitat into many niches occupied by a series of relatively similar, dentally progressive mammals. The much more rapid and extensive evolution of the characters diagnostic of the living members of the lineages must have been triggered later by a substantially larger environmental perturbation which created radically new ecological opportunities for mammals immediately after the end of the Cretaceous.

The case of the placental mammals is important because it raises the questions of why the morphology failed adequately to reveal their phylogenetic interrelationships, why the fossil record fails to correspond to the molecular-based estimates of the dates of divergence, and what is the relationship between the supraordinal groupings and the historical geology of the continental masses, which latter is actually a more accurate predictor of the main superorders than is morphology.

However, its importance extends further because the lessons to be learned from the molecular taxonomy of placental mammals can be applied by analogy to fossil taxa that lack living members. Where the cladogram of such a taxon is based on a level of morphological support that is no greater than that for the old, discredited placental mammal cladogram, then its accuracy must be in doubt. The response to such a suggestion may be that, in the absence of molecular data, the best supported morphological cladogram is the best hypothesis possible, however weak that support may be, and that there is no way of improving on it. There are, though, other possible sources of phylogenetically relevant information that have not always been as extensively explored as they deserve. One is palaeobiogeography, which relates branching points in the phylogeny to vicariant or dispersal events that can sometimes be correlated with tectonic movements of land masses, or opening up of potential dispersal routes, as revealed by the geological record (e.g. Upchurch et al., 2002). Another potentially independent source of phylogenetic information is functional analysis. A given cladogram implies a particular sequence of transformation of characters between the nodes, which themselves can be taken to represent the sequence of hypothetical ancestors and descendants. In so far as each such hypothetical ancestor had to be a fully integrated, functionally coherent phenotype, then one cladogram may be judged better than another because it implies a functionally more plausible sequence of evolutionary transitions (Kemp, 1988). To give an example, different authors have produced very different phylogenies of the therapsids of the Late Permian. Focussing on the Anomodontia that were mentioned earlier, this therapsid taxon has been claimed to be the sister group of, respectively, Dinocephalia, Therocephalia, Eutheriodontia (Therocephalia+Cynodontia), and Theriodontia (Gorgonopsia+Eutheriodontia) (Kemp, 2009). In every case, the proposed synapomorphies supporting the respective relationship are few and mostly trivial or poorly defined. However, if the sequence of increasingly derived basal members of the Dicynodontia (Fig. 1.2) are interpreted functionally in terms of increasingly modified adaptations of the skull, dentition and jaw musculature for dealing with a herbivorous diet, then the most plausible hypothetical ancestral structure resembles a generalised primitive therapsid that possessed none of the derived characters of any of those other therapsid taxa (Kemp, 2005, pp. 79–80). Therefore the functional analysis resolves the phylogenetic position of Anomodontia as a basal therapsid divergence, unrelated to any of the other groups.

1.3.2   The Molecular Revolution: Molecular Developmental Genetics

It has long been a hope of evolutionary biologists that the molecular genetic basis underlying phenotypic evolutionary change will eventually be understood in enough detail to complete the connection between genetics and evolution that began with the synthetic theory. The still very young discipline of evolutionary developmental genetics, "EvoDevo", is explicitly concerned with the causal relationship between what the fossil record and comparative morphology show to have been the course of phenotypic change and what molecular genetics shows to have been the molecular basis for it. One of the most intensively studied cases is the origin of tetrapod limb, which is particularly promising because of the combination of fossil evidence about its evolution with its role over many decades as a model system for embryology, traditional and latterly molecular (Hall, 2007). The broad features of the morphological transition from fish fin to tetrapod limb are illustrated by the sequence of Upper Devonian fossils from *Eusthenopteron* to *Ichthyostega* mentioned earlier (Shubin et al., 2006; Coates and Ruta, 2007). Meanwhile, techniques for demonstrating the timing and regions of gene expression in both normal and mutant individuals, to date mainly of mice, chick and zebrafish, are beginning to unravel the molecular basis of the development (Tanaka and Tickle, 2007). Exciting as the prospect is, however, there is a very long way to go before there are comprehensive hypotheses about how mutations in particular genes caused particular aspects of the transition from fin to limb. A bewilderingly large number of genes and gene products have been shown to act in the overall development of the vertebrate paired appendages (e.g. Arias and Stewart, 2002; Tanaka and Tickle, 2007). At present there is little agreement even on the fundamental question of whether the tetrapod autopodium (hand and foot) evolved by modification of pre-existing structures in the fish fin, or is neomorphic (Wagner and Larsson, 2007). It is possible only very tentatively to suggest certain steps that might have occurred in the sequence of genetic evolution, based on simple comparisons between modern fish and tetrapods. Wagner and Larsson (2007), for example, have recently proposed that a general autopodial developmental module evolved as a consequence of the separation of the domains of expression of two homeobox genes, Hoxa 11 and Hoxa 13. Subsequent evolution of digits within this new autopodial field was related to the acquisition of new functions by HoxD genes (e.g. Kmita et al., 2002). At some point the genes known to be involved in the identity of specific digits, such as the Shh (sonic hedgehog) and Gli3, were recruited into the system.

A second potentially illuminating example concerning late Palaeozoic and Mesozoic tetrapods is that of the synapsid jaw. Depew and his colleagues (Depew et al., 2005; Depew and Simpson, 2006) have studied the expression of genes involved in patterning of the mandibular arch, and the effects of their mutations in mutant mice. In attempting to account for different proportions of the elements of the mandibular arch amongst vertebrates, and the maintenance of functional integration between maxillary and mandibular components, they propose a "hinge and caps" model. The hinge region is presumed to be the first source of positional infor-mation for the developing mandibular arch, and integration of signals from the hinge and from their own respective attachments leads to the correct registration of upper and lower jaws. As in the limb, a very large number of genes are expressed dur-ing mandibular development, and presumably variations in the timing, position and strength of their expressions is responsible for variation in the morphology amongst different vertebrates. In the case of mammals, reduction of the posterior component of the maxillary and mandibular elements as the hinge bones gradually reduced and eventually converted into ear ossicles is assumed to have been one such outcome. The hope is that eventually it will be possible to hypothesise just what sequence of genetic mutations caused this condition, but again evolutionary developmental biology is a long way off this goal.

Indeed, given the complexity of the network of integrated gene activity involved in these examples, it is not even certain that the information available from the fos-sil record will ever have the resolution to test hypotheses about the genetic basis of evolutionary transition. Undoubtedly however, there is a great deal yet to be learned about the relationship between genotype and phenotype that will bear on the question.

1.3.3 Computed Tomography and Finite Element Analysis

A serendipitous consequence of the widespread introduction of CT scanning in medicine has been the availability of equipment for scanning fossil material. With varying degrees of resolution, a fossil can effectively be non-invasively sectioned, visually reconstructed in three dimensions, and the reconstruction can even be cor-rected for post-mortem damage and distortion. As use of the technique spreads, a rapid increase in anatomical knowledge can be expected, complete with comput-erised descriptions and atomisation into characters for multivariate and phylogenetic analysis.

CT scanning also lends itself to the application of engineering techniques for analysing mechanical structure. Finite element analysis (FEA) is a computational method for visualising the patterns of stress and strain in a structure that is sub-jected to a regime of applied forces. It is beginning to be used in palaeobiology to investigate the stresses generated within skeletal elements by the estimated forces of the reconstructed muscles. It is then possible to relate the biological design of the anatomical structure to its mechanical function as a transmission system for the stresses generated by feeding, locomotion etc. Rayfield et al. (2001) applied the method to the jaw mechanics of the large theropod dinosaur *Allosaurus*. By revealing the pattern of stresses within the reconstructed cranium induced by the action of inferred jaw muscles during biting, they could demonstrate the relation-ship between aspects of the skull design and the force pattern. For example, they showed that the large antorbital fenestrae in the skull do not weaken it significantly, because the bars of bone surrounding them act as compressive struts, effectively distributing the stress between the maxillae and the robust skull roof. In a later study, Rayfield (2005) compared the skull mechanics of three different theropods, *Coelophysis*, *Allosaurus* and *Tyrannosaurus*, and showed that aspects of the morphological differences between them correlate with differences in the stress patterns. In the first two forms, the fronto-parietal region of the skull roof is strongly built and it is in this region that the compressive and shear stresses peak. In contrast, the nasal region of *Tyrannosaurus* receives the highest stresses, and here it is this part that is the more robust region. Presumably the differences reflect different diets and modes of jaw use during feeding. As with all phenotypic differences, these may be mapped onto a phylogeny of the theropods in order to generate hypotheses of the functional significance of the inferred evolutionary divergencies in cranial anatomy (Barrett and Rayfield, 2006).

There are other aspects of tetrapod evolution that are amenable to FEA analysis, though always bearing in mind Alexander's (2006) caution about the extent of the uncertainty about the anatomy and properties of the soft tissue components in fossil vertebrates. By applying the assumed locomotory muscle forces to a 3D reconstruction of the limb of a tetrapod that is placed in a variety of possible orientations relative to the ground, it will be possible to discover which posture and gait minimises the stresses generated in the bones, with the implication that these reflect the animal's normal mode of locomotion in life. Again, as with dinosaur cranial mechanics, the technique will illuminate the functional significance of transitions to radically new modes of locomotion, for example the origin the bipedality of dinosaurs (Hutchinson, 2004), parasagittal gait of mammals (Kemp, 1978), and the flight of birds (Garner et al., 1999; Clarke et al., 2006) and pterosaurs (Wilkinson, 2007).

Another example is the application of FEA by Srivastava et al. (2005) to dinosaur eggshells, where comparison with birds' eggs suggests that thin-shelled species such as *Megaloolithus jabalpurensis*, were adapted for more arid conditions. They were able to relate the magnitude of the stresses on the egg shell to its microstructure in different species. Those with thinner shells experience higher stresses, as would be expected, but thinness is also correlated with the presence of additional subspherolith elements in the structure, which compensates by increasing the strength.

1.3.4 New Techniques for Analysing the Geochemical Record

Methods for measuring extremely small quantities of rare stable isotopes and trace elements have revolutionised the study of the palaeoenvironmental setting of fossils, and the search for the causes of the great events in the history of the Earth's biota. Of no part of the fossil record is this more true than that of the Late Palaeozoic and Mesozoic, during which the evolution and diversification of tetrapods was intimately tied up with four of the "big five" mass extinction events of the Phanerozoic. The Late Devonian crisis occurred around the time when the tetrapods originated, while the Mesozoic itself is, of course, bounded by the end-Permian and end-Cretaceous events that so affected tetrapod history. In between these two, the late Triassic event occurred around the time of the extinction of several archosaur, synapsid, and rhynchosaur taxa, the origin of the mammals, and the beginning of the great dinosaur radiation.

In particular, the last couple of decades have witnessed the development of techniques for estimating several potentially critical environmental parameters, such as palaeotemperatures on the basis of O_2 isotope ratios. There are a number of methods for estimating atmospheric CO_2 levels, including boron and carbon isotopes, and calculated volumes of buried organic carbon in palaeosols (Royer et al., 2004), and this also gives an indirect guide to palaeotemperatures on the basis of the greenhouse effect. The atmospheric O_2 level can be measured by the extent of sulphur, bacterial and therefore anaerobic activity determined from sulphur isotopes, among other methods (Berner et al., 2000). The severity of continental weathering and therefore aspects of the climate are indicated by measures of strontium isotopes. Biologically, photosynthesising organisms preferentially fix ^{13}C over ^{12}C, so the proportions of these isotopes in fossil marine shells give an indirect measure of the primary productivity. The ratio of these isotopes also differs in different kinds of plants, and therefore analysing the enamel of teeth can give a clue to the diet of herbivorous tetrapods.

As an example of the way in which this more detailed palaeoenvironmental evidence may help account for significant evolutionary events, Kemp (2006) considered the conditions of the mid-Permian, the time when the basal "pelycosaur-grade" synapsids were replaced by the more progressive early therapsids. There is no evidence for a major environmental perturbation such as a mass extinction event, but the geochemical indicators do reveal a period in which the temperature had been gradually rising from the level during the Permo-Carboniferous glaciation to about 3°C higher than today. The estimated O_2 level, though declining, was still 27% above modern levels, and the CO_2 level had risen to about three times the present-day value. He proposed a model in which the origin of the therapsids, a taxon whose morphology indicates substantially higher metabolic rates and activity levels than "pelycosaurs", was correlated with the higher oxygen availability. This higher energy budget was itself associated with the evolution of physiological regulatory mechanisms that adapted therapsids for maintaining their activity throughout the increased seasonality brought on by the rising global temperature.

1.4 New Concepts

New information, whether from discovering new fossils or from applying new methods to existing material, leads to more detailed answers to the questions about long-term patterns and processes of evolutionary change and the environmental conditions under which they occurred. This continual process of development is occasionally accompanied by the spread of a radical new concept, or way of thinking about the major evolutionary events that are illustrated by the fossil record. For the last half a century, most of the palaeobiological interpretation of fossils and its associated stratigraphic information has been dependent on a pair of simplifying concepts. The first is in a general sense atomism (Rieppel, 2001), where it is assumed that an organism consists of many discrete, more or less mutually independent characters. This assumption was found to be necessary for tractable phylogenetic analysis, and remains as important for modern computerised cladistic methodology as it ever was for traditional, non-mathematical systematic methods. Furthermore, it allows evolutionary change to be described and accounted for solely in terms of morphological shifts in discrete, identifiable traits.

The second simplifying concept is reductionism, in which it is assumed that evolutionary changes, even major morphological transitions, are caused by a simple, potentially identifiable natural selection force acting on the lineage of successive phenotypes. This is the widely accepted and rarely disputed view that the major evolutionary events of macroevolution are caused by no more than extrapolation of population level processes of microevolution acting for long enough. Therefore all that is required to account for a particular evolutionary transformation is that the one single dimension of the environment guiding the direction of the evolutionary change be identified.

Comparable atomistic and reductionist concepts have also been applied to much of the study of the palaeoecological background to major evolutionary events, such as mass extinctions and explosive radiations, and the rise of radically new kinds of community. It is usually assumed that in effect the environment consists of more or less independent parameters, and that a perturbation of one of these alone can be the cause of some great event. As far as palaeocommunity structure and dynamics are concerned, the reductionist assumption allows the processes known from studies of modern ecology, such as interspecific competition, population regulation mechanisms by predator-prey interactions, etc., can be offered as the sole cause of even those major changes that are only revealed on the geological time scale.

Of course, from Aristotle through Goethe onwards there has never been a shortage of critics of the simplification inherent in these twin concepts, and in more recent years they have been represented by the writings of, for example, Dullemeijer (1974, 1980), Riedl (1977, 1978), and Gould (2002). Nor has there been an absence of apologists for atomism and reductionism pointing out, quite reasonably, that however much atomism and reductionism may simplify the real world, they do actually provide a framework for testable hypotheses – they work. More complex "scenarios" are claimed to be effectively untestable, because a model based on a more realistically large number of variables rapidly descends into chaotic behaviour, and therefore explains nothing.

The conceptual shift that is presently spreading into palaeobiology is actually yet another consequence of the molecular revolution in biology, namely what has come to be termed "systems biology" (Kirschner, 2005; Konopka, 2007). Once it became clear to molecular biologists that cellular control mechanisms, and genetic developmental modules consist of confusingly large numbers of different interacting molecules, it was obvious that neither the atomistic assumption that each molecular species can be treated as an independent entity, nor the reductionist assumption that the action of each different molecule can be predicted from its structure alone could explain these cellular-level processes. Rather, it is the nature of the interactions between the many different molecules that determine the properties of the integrated system as a whole. For a long time engineers have used a systems approach to such things as control mechanisms for highly complex machinery, and molecular biology is adopting the same general methods (Ceste and Doyle, 2002). Such interactive phenomena within networks as signalling, feedback, inhibition, synergy, parallel pathways, and so on are more useful for explaining the properties of a system than merely describing the nature of the interacting entities themselves.

The current shift in the direction of a systems approach to palaeobiology may be illustrated by two different areas of investigation.

1.4.1 The Origin of Major New Taxa: Correlated Progression

One of the ultimate quests of evolutionary biology is surely elucidation of the mechanism by which an evolving lineage undergoes the kind of long trek through morphospace that involves large changes in numerous traits, and that therefore culminates in a radically new kind of organism – a new higher taxon. It is therefore surprising how little attention has been paid to this problem by the evolutionary biology community at large. As explained, the inhibition is primarily due to the atomistic and reductionist concepts. These underwrite a model of evolution in which even the most extensive of evolutionary change is due to "normal", that is to say microevolution, driven by natural selection acting on one, or at most perhaps two or three characters at a time within an interbreeding population. Corollaries of this model include the concept of key innovation – the idea that an evolutionary modification of some particular "key" character on its own opens up a new adaptive zone, and the familiar idea of preadaptation – which again attributes special significance to certain specified characters. From the environmental perspective, the model assumes that there was a relatively very simple selective force involved, such as for a new food source, a more effective means of escaping predators by increasing running speed, or whatever.

Such a simple view of how lineages and their characters undergo large evolutionary transformation is manifestly unrealistic. The phenotypic characters of an organism are certainly not independent of one another, but are structurally and functionally inter-dependent parts of a highly integrated system. Nor does natural selection actually act on individual characters, but on organisms as a whole, for fitness is a property of an organism that result from the integrated action of all its traits. At times it may appear that the available variation of some particular character has a more critical effect on an organism's fitness than the variation in others. However, the actual fitness of an individual bearing such a favourable character still depends on the integrated relationship of this character to many other characters within the organism.

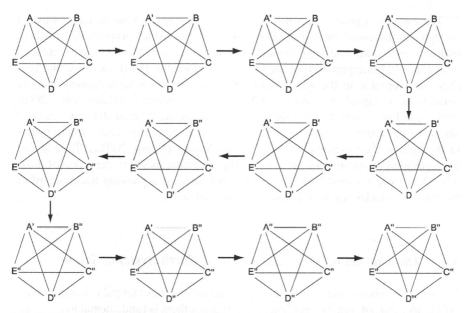

Fig. 1.5 The correlated progression model, illustrated by a five-trait phenotype in which all the traits are functionally interlinked. No more than a small incremental change in any one trait, such as A′ to A″, is possible unless and until correlated small changes in the others have occurred (from Kemp, 2007b)

An alternative model (Fig. 1.5) that is much more realistic can be derived from the concept of correlated progression (Thomson, 1966; Kemp, 2007a, b). The main assumption is that all the traits are functionally interlinked in such a fashion that the phenotype acts as an integrated system, but that there is a sufficient degree of flexibility in the functional and structural connections between traits that any one of them can change to a small extent, without losing its integration within the phenotype as a whole. No further change in that trait is presumed possible unless and until appropriate, comparably small changes have occurred in all the other traits to which it is functionally connected: in this way the integration of the phenotype is maintained as the lineage traverses even very long distances through morphospace.

The model also has implications for the nature of the selection force driving long term evolutionary change, which is assumed to act on the phenotype as a whole, and not on individual, atomised traits. Therefore, over significant evolutionary time, the selection force must be regarded as a complex of many ecological parameters: indeed, in principle all the parameters that affect the organism's life. In consequence, a long-term evolutionary trend from an ancestral to a highly derived phenotype results from the evolving lineage tracking a very general ecological gradient, rather than as a response to any single identifiable aspect of it. In essence, the correlated progression model is a systems approach, because it is based on the nature of the integrated interactions between the parts of the organism, rather than on the nature of the individual parts as such.

The correlated progression model is particularly appropriate in the context of interpreting the evolution of Late Palaeozoic and Mesozoic tetrapods, for it is here that a fossil record implying sequences of acquisition of traits can be combined with a detailed interpretation of the functional significance of those traits, and how they are integrated in the whole organism. The model has been applied in some detail to the synapsid fossil record and the origin of mammals (Kemp, 1985, 2005, 2007b), and in outline to the origin of tetrapods in the light of the new evidence about the sequence of acquisition of tetrapod characters (Fig. 1.1b), and to turtles in the light of their extremely modified morphology (Kemp, 2007a). Time is ripe for an explicit application of the correlated progression model to the origin of those other Mesozoic tetrapod higher taxa for which there is a growing fossil record of intermediate grades, such as dinosaurs, birds, and snakes.

1.4.2 The Causes of Mass Extinctions: Earth Systems Science

Mass extinctions are one of the most important discoveries that palaeontogists have ever made, and solving the problem of what causes them is fundamental to earth science. It is also fundamental to evolutionary biology because of the dramatic effect these crises had on the course of the history of life on Earth. Most of the past literature on the subject has been predicated upon a belief that mass extinctions are caused by a relatively simple, single change in the environment, one that required little imagination to see how it would devastate a wide swathe of different kinds of organisms. For example, different authors have attributed the end-Permian mass extinction event to, respectively, nutrient collapse, a bolide impact, high CO_2 level and a greenhouse effect, anoxia, volcanism, methane extrusion, and hydrogen sulphide (Bambach, 2006). However, such single-trigger models, even allowing for a cascade of subsequent effects, fail to account adequately for all the geochemical and geophysical signals associated with the event. Certainly there is some supporting evidence for each of these proposed causes (Hallam and Wignall, 1997; Erwin and Jin, 2002; Benton, 2003; Bambach, 2006; Twitchett, 2006). Reduced primary productivity is indicated by carbon isotope ratios; massive volcanism by the formation of the Siberian Traps; a bolide impact by the geochemistry of deposits in Meishan, China; anoxia by shifts in stable isotope ratios of carbon and sulphur, and also black shale deposits; temperature increase by a shift in oxygen isotope ratios and the nature of preserved terrestrial palaeosols; high methane levels by the magnitude of the carbon isotope ratio shift; a rise in CO_2 by various of these signals. In addition to these, there is stratigraphic evidence for a major regression of the sea, followed by a rapid transgression, and for active tectonic events as Pangaea was commencing its break up. Because of the low temporal resolution of the stratigraphic record, and the globally dispersed occurrence of strata, there is also a highly incomplete biological picture of the time-course of mass-extinctions (Bambach, 2006). From start to completion of the end-Permian event may have taken anywhere between the order of 10^{-2} (i.e. days) and 10^6 years and may have been a single catastrophic, a gradual, or a stepped process, and yet still appear in fossil record to have been an instantaneous event. Whether all the geochemical, stratigraphic, and biotic signals are contemporaneous or sequential is not even determinable.

The second major mass extinction associated with the Mesozoic occurred in the Late Triassic. It is associated with a comparable plethora of abiotic signals, and is therefore as shrouded in mystery as to its timing, course, and cause as is the end-Permian event (Tanner et al., 2004).

The observation that all mass extinctions are accompanied by a considerable variety of different signals, and that no two mass extinction events are ever associated with exactly the same combination of such signals is not of course new. A number of authors have proposed flow-diagrams for particular cases to illustrate possible interrelationships between the different environmental perturbations, and how these might have affected the biota. However, these have all tended to be based on the assumption that there was a potentially identifiable, single trigger that led, directly or indirectly, to a cascade of secondary effects. For the end-Permian event, Hallam and Wignall (1997) and Wignall (2001) suggested that the trigger was the gaseous content of the volcanic output associated with the formation of the Siberian traps, and that this had a series of consequences (Fig. 1.6). This model certainly can account for a number of the abiotic signals, but it is not at all clear whether all of them, including those indicating sea level change, tectonic activity, or a possible bolide impact are coincidental or causally related.

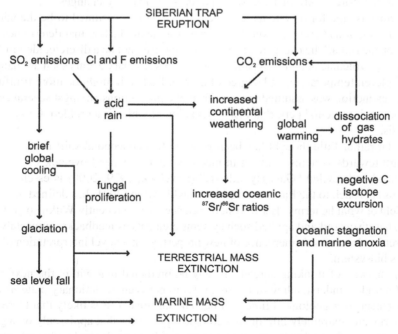

Fig. 1.6 Wignall's proposed scheme of the interrelationships of the abiotic factors associated with the end-Permian mass extinction (Redrawn from Benton, 2003)

The traditional twin concepts of reductionism and atomism, shown to be inadequate in the context of macroevolution, are similarly under challenge here. The reductionist view that the causes of extinction of species in ecological time can simply be extrapolated to geological time is difficult to sustain. Bambach (2006, quoting a personal communication from Payne and Fischer) pointed out that for a species whose population size was of the order of 10^{15}, if the death rate differed from the birth rate by only 0.1%, then the species would be extinct within 30,000 generations. For a typical fossil species, or a very large number of such species simultaneously, the extinction would appear instantaneous, because this time course is so far below the temporal resolution of the vast proportion of the fossil record. Yet a differential of such small magnitude would be impossible to detect even in a modern ecological setting. If the actual time course for the extinction event was, say, 100,000 years or more, which would still appear instantaneous, then the death rate to birth rate differential would be so minute that it is difficult to imagine an environmental perturbation that was actually small enough to be the determinate cause. In this light, the complex perturbations demonstrable during mass extinctions would seem to be vastly greater in magnitude than necessary.

The atomistic approach to the causes of mass extinction assumes that the parameters of the palaeoenvironment – temperature, levels of the different atmospheric gases, sea level, and so on – can be treated as discrete elements, as also can the various elements of the biota. Yet realistically, the various environmental and biotic elements must always have been interacting as a complex, integrated system. In this context therefore, study of large scale patterns of diversity changes such as mass extinctions is ripe for a systems approach, in which it is assumed to be the nature of the interactions between several palaeoenvironmental factors that determines the effect on the biota. The cause of a mass extinction need not be a discrete, identifiable trigger like volcanic activity, causing a cascade of secondary events such as changes in CO_2 level, temperature, and degree of anoxia. Rather, it might be more fruitful if a mass extinction was assumed to result from the interactions amongst several coincidental environmental perturbations, elucidation of which is a problem for systems analysis.

So far rather little thought has been given to this conceptual shift that parallels the shift towards systems thinking in macroevolution. James Lovelock's variously celebrated and reviled "Gaia Hypothesis" (Lovelock, 1979, 2000) is essentially a systems approach to the Earth and its biota, and Lawton (2001) has defined the general field of what he terms "Earth System Science". More recently Wilkinson (2003, 2006) has discussed and applied such systems concepts as feedbacks, autocatalysis, hierarchical levels, and emergence of new properties, in a novel interpretation of the earth's biosystem.

As this way of thinking spreads, it may be predicted that within the next couple of decades understanding of the relationship between evolutionary patterns and environment over geological time will become far greater. It is likely that there are long-term, and extremely infrequent ecological processes that apply only on a geological time-scale, and that cannot be discovered from ecological time-scale study alone. It is only in this light that such events as mass extinctions and major episodes of evolutionary radiation may eventually be understood.

1.5 Conclusion

The most radical new perspectives on the evolution of Late Palaeozoic and Mesozoic terrestrial tetrapods are indirectly due to the revolution in molecular biology, in which very large amounts of DNA of many modern species has been sequenced, alongside the development of powerful computer programmes and sophisticated statistical methods for its interpretation. This has thrown a great deal of new light on the evolutionary patterns and processes of living organisms, insights that can increasingly be applied by analogy to fossil organisms. Highly robust molecular-based phylogenies of taxa like birds and mammals, whose early branchings were closely spaced morphologically, have exposed the very limited extent to which morphological characters reliably reveal relationships in such cases. In turn, this is generating caution about accepting the accuracy of the purely morphological-based phylogenies of comparable extinct taxa, such as therapsids and dinosaur subgroups, and correspondingly more attention must be paid to potentially corroborating biogeographical and functional evidence of relationships.

Increasing understanding of the relationship at the molecular level between genes, developmental processes, and phenotyopic structure is leading to a greater understanding of how gene mutations in known regulatory gene families might have caused particular morphological transitions that are inferred from the fossil record of stem tetrapods, mammals and other higher taxa. This aspect of molecular biology has as yet generated much less in the way of firm palaeobiological results than the systematics aspect, but promises eventually to be at least as profound.

The third new perspective also owes its origin to the molecular biology revolution, though even more indirectly. Systems biology is being developed for analysing and understanding how complex, dynamic molecular systems work. A comparable systems approach can be applied to large scale phenotypic evolution, and to the complex environmental circumstances associated with major evolutionary events. In the near future, computer simulations of evolving sequences of characters as inferred from stem group fossils will lead to a deeper understanding of the evolutionary causes of the origin of such major taxa as tetrapods, amphibians, chelonians, snakes, dinosaurs, birds and mammals. Similarly, viewing the palaeoenvironment and its perturbations in geological time as a system of integrated parameters will allow a greater understanding of how the environment is implicated in the great evolutionary events of mass extinction, ecological replacement, explosive radiation, and origin of new higher taxa, as uniquely revealed in the fossil record.

References

Alexander, RM (2006) Dinosaur biomechanics. Proc R Soc, B273:1849–1855.

Apesteguía, S, Zaher, H (2006) A Cretaceous terrestrial snake with robust hindlimbs and a sacrum. Nature, 440:1037–1040.

Archibald, JD (2003) Timing and biogeography of the eutherian radiation: fossils and molecules compared. Mol Phylogenetics Evol, 28:350–359.

Archibald, JD, Averianov, AO, Ekdale, EG (2001) Late Cretaceous relatives of rabbits, rodents, and other extant eutherian mammals. Nature, 414:62–65.

Arias, AM, Stewart, A (2002) Molecular principles of animal development. Oxford University Press, Oxford.

Asher, RA, Novacek, MJ, Geisler, JH (2003) Relationships of endemic African mammals and their fossil relatives based on morphological and molecular evidence. J Mammalian Evol, 10:131–194.

Bambach, RK (2006) Phanerozoic biodiversity and mass extinctions. Ann Rev Earth Plane Sci, 34:127–155.

Bandyopadhyay, S, Sengupta, DP (2006) Vertebrate faunal turnover during the Triassic-Jurassic transition: an Indian scenario. N M Mus Nat Hist Sci Bull, 37:77–85.

Barrett, PM, Rayfield, EJ (2006) Ecological and evolutionary implications of dinosaur feeding behaviour. Trends Ecol Evol, 21:217–224.

Bell, MA (2000) Bridging the gap between population biology and paleobiology. Evolution, 54:1457–1461.

Benton, MJ (1994) Late Triassic to Middle Jurassic extinctions among continental tettrapods: testing the pattern. In: Fraser, NC, Sues, H-D (eds) In the shadow of the dinosaurs. Cambridge University Press, Cambridge, MA.

Benton, MJ (1996) On the nonprevalence of competitive replacement in the evolution of tetrapods. In: Jablonski, D, Erwin DH, Lipps, JH (eds) Evolutionary paleobiology. Chicago University Press, Chicago, IL.

Benton, MJ (2003) When life nearly died: the greatest extinction of all time. Thames and Hudson, London, UK.

Berner, RA, Petsch, ST, Lake, JA, Beerling, DJ, Popp, BN, Lane, RS, Laws, EA, Westley, MB, Cassar, N, Woodward, FI, Quick, WP (2000) Isotope fractionation and atmospheric oxygen: implications for Phanerozoic O_2 evolutionn. Science, 287:1630–1633.

Bininda-Emonds, ORP, Cardillo, M, Jones, KE, MacPhee, RDE, Beck, RMD, Grenyer, R, Price, SA, Vos, RA, Gittleman, JL, Purvis, A (2007) The delayed rise of present-day mammals. Nature, 446:507–512.

Boisvert, CA (2005) The pelvic fin and girdle of *Panderichthys* and the origin of tetrapod locomotion. Nature, 438:1145–1147.

Butler, RJ (2005) The 'fabrosaurid' ornithischian dinosaurs of the Upper Elliot Formation (Lower Jurassic) of South Africa and Lesotho. Zool J Linn Soc, 145:175–218.

Caldwell, MW, Lee, MSY (1999) A snake with legs from the marine Cretaceous of the Middle East. Nature, 386:705–709.

Ceste, ME, Doyle, JC (2002) Reverse engineering of biological complexity. Science, 295:1664–1669.

Charlesworth, B (1996) The good fairy godmother of evolutionary genetics. Curr Biol, 6:220.

Chatterjee, S, Templin, JR (2007) Biplane wing planform and flight performance of the feathered dinosaur *Microraptor qui*. Proc Natl Acad Sci USA, 104:1576–1580.

Clack, JA (1994) *Acanthostega gunnari*, a Devonian tetrapod from Greenland: the snout, palate and ventral parts of the braincase, with a discussion of their significance. Meddelelser Gronland Geosci, 31:1–24.

Clack, JA (2002) Gaining ground: the origin and evolution of tetrapods. Indiana University Press, Bloomington, IN.

Clarke, JA, Zhou, Z, Zhang, F (2006) Insight into the evolution of avian flight from a new clade of early Cretaceous ornithurines from China and the morphology of *Yixianornis grabaui*. J Anat, 208:287–308.

Coates, MI (1996) The Devonian tetrapod *Acanthostega gunneri* Jarvik: postcranial anatomy, basal tetrapod relationships and patterns of skeletal evolution. Trans R Soc Edinburgh Earth Sci, 87:363–421.

Coates, MI, Ruta, M (2007) Skeletal changes in the transition from fins to limbs. In: Hall, BK (ed) Fins into limbs: evolution, development, and transformation. Chicago University Press, Chicago, IL.

Daeschler, EB, Shubin, NH, Jenkins, FA (2006) A Devonian tetrapod-like fish and the evolution of the tetrapod body plan. Nature, 440:757–763.

Depew, MJ, Simpson, CA (2006) 21st Century neontology and the comparative development of the vertebrate skull. J Exp Zool (Mol Dev Evol), 235:1256–1291.

Depew, MJ, Simpson, CA, Morasso, M, Rubenstein, JLR (2005) Reassessing the Dlx code: the genetic regulation of branchial arch skeletal pattern and development. J Anat, 207:501–561.

Donaghue, PCJ, Benton, MJ (2007) Rocks and clocks: calibrating the tree of life using fossils and molecules. Trends Ecol Evol, 22:424–431.

Dullemeijer, P (1974) Concepts and approaches in animal morphology. Van Gorcum, Assen, The Netherlands.

Dullemeijer, P (1980) Functional morphology and evolutionary biology. Acta Biotheor, 29:151–250.

Erwin, DH, Jin, SA (2002) End-Permian mass extinctions: a review. Geol Soc Am Sp Pap, 356:363–383.

Foote, M, Hunter, JP, Janis, C, Sepkoski, JJ (1999) Evolutionary and preservational constraints on origins of biologic groups: divergence times of eutherian mammals. Science, 283:1310–1314.

Garner, JP, Taylor, GK, Thomas, ALR (1999) On the origin of birds: the sequence of character acquisition in the evolution of bird flight. Proc R Soc, B266:1259–1266.

Gould, SJ (1994) Tempo and mode in the macroevolutionary reconstruction of Darwinism. Proc Natl Acad Sci USA, 91:6764–6771.

Gould, SJ (2002) The structure of evolutionary theory. Belknap Press, Harvard, Cambridge, MA.

Grantham, T (2007) Is macroevolution more than successive rounds of microevolution? Palaeontology, 50:75–85.

Hall, BK (2007) Fins into limbs: evolution, development, and transformation. Chicago University Press, Chicago, IL.

Hallam, A, Wignall, PB (1997) Mass extinctions and their aftermath. Oxford University Press, Oxford.

Hu, D, Hu, L, Zhang, L, Xu, X (2009) A pre-*Archaeopteryx* troodontid theropod from China with long feathers on the metatarsus. Nature, 461:640–643.

Hu, Y, Meng, J, Wang, Y, Li, C (2005) Large Mesozoic mammals fed on young dinosaurs. Nature, 433:149–152.

Hunter, JP, Janis, CM (2006) Spiny Norman in the Garden of Eden? Dispersal and early biogeography of Placentalia. J Mammal Evol, 13:89–123.

Hutchinson, JR (2004) Biomechanical modeling and sensitivity analysis of bipedal running ability. II. Extinct taxa. J Morphol, 262:441–461.

Irmis, RB, Nesbitt, SJ, Padian, K, Smith, ND, Turner, AH, Woody, D, Downs, A (2007) A Late Triassic dinosauromorph assemblage from New Mexico and the rise of dinosaurs. Science, 317:358–361.

Jarvik, E (1980) Basic structure and evolution of vertebrates, Vols. 1 and 2. Academic Press, New York, NY.

Ji, Q, Luo, Z-X, Wible, JR, Zhang, J-P, Georgi, JA (2002) The earliest known eutherian mammal. Nature, 416:816–822.

Ji, Q, Luo, Z-X, Yuan, C-X, Tabrum, AR (2006) A swimming mammaliaform from the Middle Jurassic and ecomorphological diversification of early mammals. Science, 311:1123–1127.

Jin, M, Yaoming, H, Chuankui, L, Yuanqing, W. (2006) The mammal fauna in the Early Cretaceous Jehol Biota: implications for diversity and biology of Mesozoic mammals. Geol J, 41:439–463.

Kemp, TS (1978) Stance and gait in the hindlimb of a therocephalian mammal-like reptile. J Zool London, 186:143–161.

Kemp, TS (1985) Synapsid reptiles and the origin of higher taxa. Spec Pap Palaeontol, 33:175–184.
Kemp, TS (1988) Haemothermia or Archosauria? The interrelationships of mammals, birds and crocodiles. Zool J Linn Soc, 92:67–104.
Kemp, TS (1999) Fossils and evolution. Oxford University Press, Oxford.
Kemp, TS (2005) The origin and evolution of mammals. Oxford University Press, Oxford.
Kemp, TS (2006) The origin and early radiation of the therapsid mammal-like reptiles: a palaeobiological hypothesis. J Evol Biol, 19:1231–1247.
Kemp, TS (2007a) The concept of correlated progression as the basis of a model for the evolutionary origin of major new taxa. Proc R Soc, B274:1667–1673.
Kemp, TS (2007b) The origin of higher taxa: macroevolutionary processes, and the case of the mammals. Acta Zool, 88:3–22.
Kemp, TS (2009) Phylogenetic interrelationships and pattern of evolution of the therapsids: testing for polytomy. Palaeontol Afr, 44:1–12.
Kielan-Jaworowska, Z, Cifella, RL, Luo, Z-X (2004) Mammals from the age of dinosaurs: origins, evolution, and structure. Columbia University Press, New York, NY.
Kirschner, MW (2005) The meaning of systems biology. Cell, 121:503–504.
Kmita, M, Fraudeau, N, Hérault, Y, Duboule, D (2002) Serial deletions and duplications suggest a mechanism for the collinearity of Hoxd genes in limbs. Nature, 420:145–150.
Konopka, AK (2007) Basic concepts of systems biology. In: Konopka, AK (ed) Systems biology: principles, methods, and concepts. CRC Press, Taylor and Francis Group, Boca Raton, FL.
Langer, MC, Benton, MJ (2006) Early dinosaurs: a phylogenetic study. J Syst Palaeontol, 4:309–358.
Lawton, J (2001) Earth systems science. Science, 292:1965.
Li, C, Wu, X-C, Reippel, O, Wang, L-T, Zhao, L-J (2008) An ancestral turtle from the late Triassic of southwestern China. Nature, 456:497–500.
Lovelock, J (1979) Gaia. Oxford University Press, Oxford.
Lovelock, J (2000) Homage to Gaia. Oxford University Press, Oxford.
Lucas, SG, Tanner, LH (2006) Tetrapod biostratigraphy and biochronology of the Triassic-Jurassic transition on the Southern Colorado plateau, USA. Palaeogeogr Palaeoclimatol Palaeoecol, 244:242–256.
Luo, Z-X, Ji, Q, Wible, JR, Yuan, C-X (2003) An Early Cretaceous tribosphenic mammal and metatherian evolution. Science, 302:1934–1940.
Luo, Z-X, Wible, JR (2005) A Late Jurassic digging mammal and early mammalian diversification. Science, 308:103–107.
McKenna, MC, Bell, SK (1997) Classification of mammals above the species level. Columbia University Press, New York, NY.
Meng, J, Hu, Y, Wang, Y, Wang, X, Li, C (2006) A Mesozoic gliding mammal from northeastern China. Nature, 444:889–893.
Modesto, SP, Rubidge, BS, Welman, J (1999) The most basal anomodont therapsid and the primacy of Gondwana in the evolution of anomodonts. Proc R Soc, B266:331–337.
Murphy, WJ, Eizirik, E, (2009) Placental mammals. In: Hedges, SB, Kumar, S (eds) The timetree of life. Oxford University Press, Oxford.
Novacek, MJ, Wyss, AR, McKenna, MC (1988) The major groups of eutherian mammals. In: Benton, MJ (ed) The phylogeny and classification of tetrapods, Vol. 2. Oxford University Press, Oxford.
Padian, K, Dial, KP (2005) Origin of flight: could the 'four-winged' dinosaur fly? Nature, 438:3–4.
Rauhut, OWM (2003) Interrelationships and evolution of basal theropod dinosaurs. Spec Pap Palaeontol, 69:3–213.
Rayfield, EJ (2005) Aspects of comparative cranial mechanics in the theropod dinosaurs Coelophysis, Allosaurus and Tyrannosaurus. Zool J Linn Soc, 144:309–316.
Rayfield, EJ, Norman, DB, Horner, CC, Hirner, JR, Smith, PM, Thomason, JJ, Upchurch, P (2001) Cranial design and function in a large theropod dinosaur. Nature, 409: 1033–1037.

Reisz, RR, Sues, H-D (2000) Herbivory in late Paleozoic and Triassic terrestrial vertebrates. In: Sues, H-D (ed) Evolution of herbivory in terrestrial vertebrates: perspectives from the fossil record. Cambridge University Press, Cambridge, MA.

Riedl, R (1977) A systems-analytical approach to macro-evolutionary phenomena. Q Rev Biol, 522:351–370.

Riedl, R (1978) Order in living organisms: a systems analysis of evolution. Wiley, Chichester.

Rieppel, O (2001) Preformationist and epigenetic biases in the history of the morphological character concept. In: Wagner, GP (ed) The character concept in evolutionary biology. Academic Press, San Diego, CA.

Rose, KD (2006) The beginning of the age of mammals. John Hopkins University Press, Baltimore, MD.

Rosenzweig, ML, McCord, RD (1991) Incumbent replacement: evidence for long-term evolutionary progress. Paleobiology, 17:202–213.

Royer, DL, Berner, RA, Montañez, IP, Tabor, NJ, Beerling, DJ (2004) CO_2 as a primary driver of Phanerozoic climate. GSA Today, 14:4–10.

Rubidge, BS (1990) Redescription of the cranial morphology of *Eodicynodon oosthuizeni* (Therapsida: Dicynodontia). Navorsinge Nasl Mus Bloemfontein, 7:1–25.

Rubidge, BS, Hopson, JA (1996) A primitive anomodont therapsid from the base of the Beaufort Group (Upper Permian) of South Africa. Zool J Linn Soc, 117:115–139.

Rybczynski, N (2000) Cranial anatomy and phylogenetic position of *Suminia getmanovi*, a basal anomodont (Amniota: Therapsida) from the Late Permian of Eastern Europe. Zool J Linn Soc, 130:329–373.

Save Soderbergh, G (1932) Preliminary note on Devonian stegocephalians from East Greenland. Meddr Gronland Geosci, 98:1–211.

Sereno, PC, Wilson, JA, Witmer, L, Whitlock, J, Maga, A, Ide, O, Rowe, T (2007) Structural extremes in a Cretaceous dinosaur. Public Lib Sci, 2:e1230.

Shubin, NH, Daeschler, EB, Jenkins, FAJ (2006) The pectoral fin of *Tiktaalik roseae* and the origin of the tetrapod limb. Nature, 440:764–771.

Springer, MS, Murphy, WJ, Eizirik, E, O'Brien, SJ (2003) Placental mammal diversification and the Cretaceous-Tertiary boundary. Proc Natl Acad Sci USA, 100: 1056–1061.

Springer, MS, Murphy, WJ, Eizirik, E, O'Brien, SJ (2005) Molecular evidence for the major placental clades. In: Rose, KD, Archibald, JD (eds) The rise of the placental mammals. John Hopkins University Press, Baltimore, MD.

Srivastava, R, Sahni, A, Jafar, SA, Mishra, S (2005) Microstructure-dictated resistance properties of some Indian dinosaur eggshells: finite element modelling. Paleobiology, 31: 315–323.

Tanaka, M, Tickle, C (2007) The development of fins and limbs. In: Hall, BK (ed) Fins into limbs: evolution, development, and transformation. Chicago University Press, Chicago, IL.

Tanner, LH, Lucas, SG, Chapman, MG (2004) Assessing the record and causes of Late Triassic extinctions. Earth Sci Rev, 65:103–139.

Thomson, KS (1966) The evolution of the tetrapod middle ear in the rhipidistian-amphibian transition. Am Zool, 6:379–397.

Turner, AH, Pol, D, Clarke, JA, Eticleson, GM, Norell, MA (2007) A basal dromaeosaurid and size evolution preceding avian flight. Science, 317:1378–1381.

Twitchett, RJ (2006) The palaeoclimatology, palaeoecology and palaeoenvironmental analysis of mass extinction events. Palaeogeogr Palaeoclimatol Palaeoecol, 232: 190–213.

Upchurch, P, Barrett, PM, Galton, PM (2007) A phylogenetic analysis of basal sauropodomorph dinosaurs. Spec Pap Palaeontology, 77:57–90.

Upchurch, P, Hunn, CA, Norman, DB (2002) An analysis of dinosaurian biogeography: evidence for the existence of vicariance and dispersal patterns caused by geological events. Proc R Soc, B269:613–621.

Vorobyeva, EI, Schultze, H-P (1991) Description and systematics of panderichthyid fishes with comments on their relationship to tetrapods. In: Schultze, H-P, Trueb, L (eds) Origins of higher groups of tetrapods. Cornell University Press, Ithaca, NY.

Wagner, GP, Larsson, HCE (2007) Fins and limbs in the study of evolutionary novelties. In: Hall, BK (ed) Fins into limbs: evolution, development, and transformation. Chicago University Press, Chicago, IL.

Wignall, PB (2001) Large igneous provinces and mass extinctions. Earth Sci Rev, 53:1–33.

Wilkinson, DM (2003) The fundamental processes in ecology: a thought experiment on extraterrestrial biospheres. Biol Rev, 78:171–179.

Wilkinson, DM (2006) Fundamental processes in ecology: an earth systems approach. Oxford University Press, Oxford.

Wilkinson, MT (2007) Sailing the skies: the improbable aeronautical success of the pterosaurs. J Exp Biol, 210:1663–1671.

Xu, X, Norrell, MA (2006) Non-avian dinosaurs from the Lower Cretaceous Jehol group of western Liaoning, China. Geol J, 41:419–437.

Xu, X, Tan, Q, Wang, J, Zhao, X, Tan, L (2007) A gigantic bird-like dinosaur from the Late Cretaceous of China. Nature, 447:844–847.

Zhang, Z (2006) Evolutionary radiation of the Jehol Biota: chronological and ecological perspectives. Geol J, 41:377–393.

Zhou, Z (2004) The origin and early evolution of birds: discoveries, disputes, and perspectives from fossil evidence. Naturwissenschaften, 91:455–471.

Zhou, Z, Barrett, PM, Hilton, J (2003) An exceptionally preserved Lower Cretaceous ecosystem. Nature, 421:807–813.

Chapter 2
The Origin, Early History and Diversification of Lepidosauromorph Reptiles

Susan E. Evans and Marc E.H. Jones

2.1 Introduction

Lepidosauria was erected by Romer (1956) to encompass diapsids that lacked diagnostic archosaurian characters. The resulting assemblage was paraphyletic. In the intervening 50 years, new fossils and new phylogenetic approaches have transformed our concepts (e.g., Evans, 1980, 1984, 1988; Benton, 1985; Whiteside, 1986; Gauthier et al., 1988). Lepidosauria is now restricted to the last common ancestor of Squamata (lizards, snakes and amphisbaenians) and Rhynchocephalia (represented by *Sphenodon*), and all descendants of that ancestor (e.g., Gauthier et al., 1988). The clade is robustly diagnosed by hard and soft characters (e.g., Gauthier et al., 1988; de Braga and Rieppel, 1997; Evans, 2003; Hill, 2005) and is recognized by recent molecular phylogenies (e.g., Gorr et al., 1998; Rest et al., 2003; Townsend et al., 2004; Vidal and Hedges, 2005). Extant lepidosaurs are globally distributed with more than 7,000 species ranging from desert lizards to marine snakes. The fossil record provides evidence of their history and radiation but, despite advances, that record is patchy. It relies mainly on microvertebrate assemblages, supplemented by rare skeletons from lacustrine and other fine grained deposits. Inevitably, the record is geographically and geologically biased.

With reanalysis, many of Romer's "lepidosaurs" were transferred to the archosaurian stem within a new clade, Archosauromorpha (Gauthier et al., 1988). These included "Prolacertiformes" (probably paraphyletic, e.g., Dilkes, 1998), Trilophosauria, and Rhynchosauria (Benton, 1985; Gauthier et al., 1988). A sister group, Lepidosauromorpha, was erected for Lepidosauria and all taxa sharing a more recent common ancestor with it than with Archosauria (Gauthier et al., 1988). At first, the clade encompassed the Permo-Triassic Gondwanan Younginiformes

S.E. Evans (✉)
Research Department of Cell and Developmental Biology, University College London, London WC1E 6BT, UK
e-mail: ucgasue@ucl.ac.uk

S. Bandyopadhyay (ed.), *New Aspects of Mesozoic Biodiversity*,
Lecture Notes in Earth Sciences 132, DOI 10.1007/978-3-642-10311-7_2,
© Springer-Verlag Berlin Heidelberg 2010

(e.g., Benton, 1985; Evans, 1988), the Triassic Laurasian Kuehneosauridae (Robinson, 1962; Colbert, 1966), and several Permo-Triassic taxa from South Africa (Carroll, 1975), but younginiforms were subsequently removed (Laurin, 1991). Currently, Archosauromorpha + Lepidosauromorpha constitute the Sauria, and Sauria + Younginiformes constitute Neodiapsida (Laurin, 1991). In a reexamination of reptile relationships, de Braga and Rieppel (1997) obtained a clade encompassing turtles, sauropterygians and lepidosaurs. Under the stem-based definition of Gauthier et al. (1988), all three groups would fall within Lepidosauromorpha. This arrangement received mixed support from Hill (2005), who recovered a weakly supported lepidosaur-turtle clade, but sauropterygians fell outside Sauria. Alternative morphological (Müller, 2004) and molecular (e.g., Hedges and Poling, 1999; Rest et al., 2003) analyses suggest that if turtles are diapsids, they are closer to archosaurs than to lepidosaurs.

2.2 The Lepidosauromorph Record

2.2.1 Permo-Triassic Lepidosauromorphs

The earliest putative lepidosauromorphs are Late Permian in age: *Lanthanolania* (Russia, Modesto and Reisz, 2002) and *Saurosternon* (South Africa, Carroll, 1975). The first is a partial skull with an incomplete lower temporal bar but no other diagnostic features, and the second is a headless skeleton. They may be basal lepidosauromorphs, or lie lower on the saurian stem (e.g., Modesto and Reisz, 2002; Müller, 2004). Of other supposed Permo-Triassic lepidosauromorphs, *Santaisaurus* (China, Sun et al., 1992) and *Colubrifer* (South Africa, Carroll, 1982) are procolophonians (Evans, 2001); *Kadimakara* (Australia, Bartholomai, 1979) is a misinterpreted specimen of *Prolacerta* (SE pers. obs.); and *Kudnu* (Australia, Bartholomai, 1979) and *Blomosaurus* (Russia, Tatarinov, 1978) are too poorly preserved to interpret with confidence but are probably also procolophonian. *Paliguana* (Early Triassic, South Africa, Carroll, 1975) is represented by a single, damaged skull with a large, flared quadrate consistent with lepidosauromorph attribution. Roughly contemporaneous remains, referable to two distinct taxa, have recently been recovered from Early Triassic fissure infillings in Poland (Czatkowice, Borsuk-Białynicka et al., 1999). These taxa are described elsewhere (Evans and Borsuk-Białynicka, 2009; Evans, 2009) and include an early kuehneosaur (see below, Kuehneosauria) and a stem-lepidosaur. The Czatkowice deposits were formed in an arid environment with localized water bodies (Borsuk-Białynicka et al., 1999). The associated fauna includes fish; temnospondyl amphibians; the proanuran *Czatkobatrachus*; procolophonians; and several archosauromorphs. Less is known about the South African Donnybrook locality (*Paliguana*), but the general Early Triassic environment of the Karoo Basin has been described as a warm, arid floodplain with rivers, playas and lakes (e.g., Smith and Botha, 2005).

2.2.2 Kuehneosauria

Kuehneosaurs are specialized, long-ribbed gliders/parachuters known from the Early Triassic of Poland and the Late Triassic (Carnian-Rhaetian) of England and North America. The English genera are *Kuehneosaurus latus* (Emborough Quarry, Robinson, 1962) and the longer-ribbed *Kuehneosuchus latissimus* (Batscombe Quarry, Robinson, 1967a). *Icarosaurus siefkeri* is based on a single skeleton from the Newark Basin, New Jersey, USA (Colbert, 1966, 1970), but partial jaws reported from the Triassic Chinle Formation (Arizona, New Mexico, Murry, 1987) are indeterminate. The new Czatkowice taxon (Evans, 2009) shows typical kuehneosaur skull morphology, but is less specialized postcranially than younger taxa. Nonetheless, the kuehneosaur *bauplan* had clearly evolved by the Early Triassic, extending the roots of the clade into the Permian (Fig. 2.1). This brings the kuehneosaurs temporally and geographically close to another group of specialized early gliders, the coelurosauravids of England, Germany and Madagascar (Evans, 1982; Evans and Haubold, 1987), but members of the two clades are morphologically distinct (SE pers. obs.).

Robinson (1962, 1967b) interpreted kuehneosaurs as primitive squamates, but this was challenged as early lepidosaurs became better known (e.g., Evans, 1980, 1984, 1988), and the first major cladistic analysis of lepidosauromorphs (Gauthier et al., 1988) placed kuehneosaurs on the lepidosaurian stem (Lepidosauria+Kuehneosauridae = Lepidosauriformes). Müller (2004) moved kuehneosaurs to the saurian stem, as the sister group of the peculiar Late Triassic Euramerican drepanosaurs. Reanalysis of Müller's matrix, with the data for *Kuehneosaurus* corrected and the Czatkowice taxon included, returned kuehneosaurs to Lepidosauromorpha.

Kuehneosaurus and *Kuehneosuchus* lived on small, relatively dry, offshore islands, in association with pterosaurs, archosauriforms, rhynchocephalians, and rare mammals (Robinson, 1962; Fraser, 1994). The Czatkowice environment was similar (Borsuk-Bialynicka et al., 1999), but *Icarosaurus* was preserved in a lacustrine assemblage of fish, temnospondyls, a drepanosaur, and several archosauromorphs including a phytosaur (Colbert and Olsen, 2001).

2.2.3 Other Mesozoic Non-lepidosaurian Lepidosauromorphs

Other designated Mesozoic lepidosauromorphs that lie outside Lepidosauria include the Middle Triassic *Coartaredens* (England, Spencer and Storrs, 2002) and *Megachirella* (Italy, Renesto and Posenato, 2003); the Early Jurassic *Tamaulipasaurus* (Mexico, Clark and Hernandez, 1994); and the Middle-Late Jurassic *Marmoretta* (UK, Portugal, Evans, 1991). *Coartaredens* (Spencer and Storrs, 2002) is represented by partial jaws that are almost certainly procolophonian (contra Spencer and Storrs, 2002). The affinities of *Megachirella* and *Tamaulipasaurus* remain unresolved.

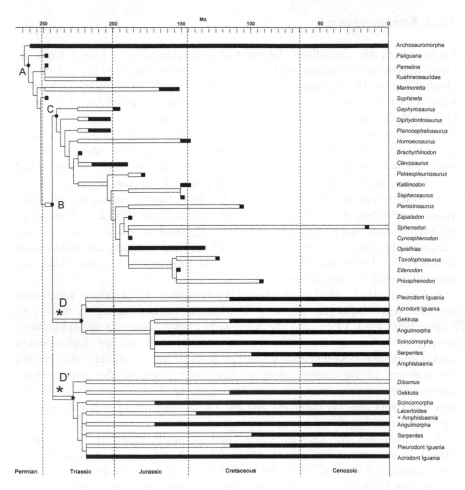

Fig. 2.1 Stratophylogenetic tree of Lepidosauromorpha using the timescale of Gradstein and Ogg (2004) with the latest possible branching points for lineages, based mainly on Evans (2003) and Jones (2006a) with additional data from Datta and Ray (2006, *Tikiguana*), Benton and Donoghue (2007, Archosauromorpha, *Protorosaurus*), Conrad and Norell (2006, Autarchoglossa [Scincomorpha+Anguimorpha], *Parviraptor*), Evans and Borsuk-Białynicka (2009, Czatkowice 1) and Evans (2009, Czatkowice 2). *A*, Lepidosauromorpha; *B*, Lepidosauria; *C*, Rhynchocephalia (phylogeny of Apesteguía and Novas, 2003); *D*, Squamata (morphological tree of Estes et al., 1988); *D'*, Squamata (modified molecular phylogeny of Townsend et al., 2004)

2.2.4 Rhynchocephalia

Rhynchocephalia (Günther, 1867) was erected for *Sphenodon* and its fossil relatives, but the later addition of unrelated acrodont taxa (e.g., rhynchosaurs, claraziids, Romer, 1956) rendered the group polyphyletic (Benton, 1985). Subsequent redefinition of a monophyletic Rhynchocephalia (Gauthier et al., 1988), based around

Sphenodon as originally intended, has now been widely accepted (e.g., Sues and Baird, 1993; Reynoso, 1996, 2000, 2005; Reynoso and Clark, 1998; Gorr et al., 1998; Lee, 1998; Ferigolo, 1999; Schwenk, 2000; Evans, 2003; Wu, 2003; Jones, 2004, 2006a, b, c). Some authors use Sphenodontida in place of Rhynchocephalia (e.g., Vidal and Hedges, 2005), but we strongly advocate usage of the original terminology.

The Early Jurassic *Gephyrosaurus* (Evans, 1980) is the sister taxon of all other rhynchocephalians, with the Late Triassic *Diphydontosaurus* (Whiteside, 1986) and the fully acrodont *Planocephalosaurus* (Fraser, 1982) crownward of it (Fraser and Benton, 1989; Wilkinson and Benton, 1996). Of remaining taxa, the Jurassic *Eilenodon* (Russmusen and Callison, 1981) and Early Cretaceous *Toxolophosaurus* (Throckmorton et al., 1981) appear related to *Sphenodon*, and a clevosaur clade is generally recognized (e.g., Reynoso and Clark, 1998), but there consensus ends (e.g., Wilkinson and Benton, 1996; Reynoso and Clark, 1998; Reynoso, 2005).

The recognition that *Sphenodon* was not a lizard (Günther, 1867) prompted a long debate as to its relationships. Its fully diapsid skull was interpreted as primitive (e.g., Watson, 1914; Parrington, 1935) and *Sphenodon* came to be regarded as a "living fossil", a surviving representative of a conservative ancient diapsid lineage. Now, the fossil record lists more than 40 rhynchocephalian taxa, with a temporal range from Late Triassic (Carnian: Scotland, Fraser and Benton, 1989; Texas, Heckert, 2004; Poland, Dzik and Sulej, 2007) to Recent, and a geographical distribution including Europe, North and South America, China, India, Morocco, South Africa, and New Zealand (Jones, 2006a, b; Jones et al., 2009). Flynn et al.'s (2006, Fig. 10) purported Middle Jurassic rhynchocephalian from Madagascar appears to be a partial theropod tooth.

Mesozoic rhynchocephalians were diverse. They ranged in size over more than an order of magnitude (Fig. 2.2) and included long-bodied marine swimmers (pleurosaurs, sapheosaurs, e.g., Carroll and Wild, 1994), gracile runners (*Homoeosaurus*, e.g., Cocude-Michel, 1963), the armoured *Pamizinsaurus* (Reynoso, 1997) and large bodied genera with hoof-like unguals (*Priosphenodon*, Apesteguia and Novas, 2003). Trophically (Jones, 2006a, c), they included insectivores (e.g., *Gephryosaurus*; *Diphydontosaurus*), opportunistic "carnivores" (*Sphenodon*, e.g., Dawbin, 1962; Cree et al., 1999), supposedly venomous predators (*Sphenovipera*, Reynoso, 2005), and specialized herbivores (e.g., *Toxolophosaurus*, Throckmorton et al., 1981; *Priosphenodon*, Apesteguía and Novas, 2003). Although in some characters they are less derived than lizards (e.g., the fifth metatarsal, the inner ear), the rhynchocephalian feeding apparatus is sophisticated (Jones, 2006a, c; Jones, 2008) and some apparently primitive traits are secondary (e.g., lack of an eardrum and quadrate conch, complete lower temporal bar). *Sphenodon* can remain active at temperatures well below those at which lizards function (5.2°C, Thompson and Daugherty, 1998). This characteristic, in concert with the long reproductive cycle and long life span (Crook, 1975), could be primitive or, more plausibly, an adaptation to life in a cool, relatively high latitude environment (Gans, 1983). The fact that New Zealand lizards show similar, though less extreme, adaptations (Gans, 1983; Cree, 1994; Bannock et al., 1999) supports the latter interpretation.

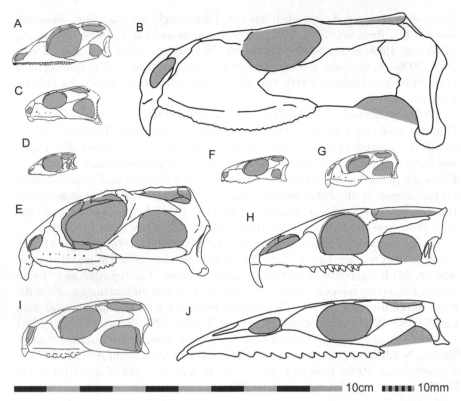

                                                                    ■■■■■ 10cm ▪▪▪▪▪ 10mm

Fig. 2.2 Lateral views of selected rhynchocephalian skulls to illustrate morphological and size diversity. (**a**) *Gephyrosaurus* (Early Jurassic, Wales, Evans, 1980); (**b**) *Priosphenodon* (Late Cretaceous, Argentina, Apesteguía and Novas, 2003); (**c**) *Brachyrhinodon* (Late Triassic, Scotland, Fraser and Benton, 1989); (**d**), *Diphydontosaurus* (Late Triassic, UK, Whiteside, 1986); (**e**) *Sphenodon* (?Miocene-Recent, New Zealand); (**f**) *Planocephalosaurus* (Late Triassic, UK, Fraser, 1982); (**g**) *Clevosaurus* (Late Triassic, Canada, Sues et al., 1994); (**h**) *Palaeopleurosaurus* (Early Jurassic, Germany, Carroll and Wild, 1994); (**i**) *Clevosaurus* (Late Triassic, UK, Fraser, 1988); (**j**) *Pleurosaurus* (Jurassic, Europe, Carroll and Wild, 1994)

2.2.5 Squamata

Squamata includes over 7,000 extant species (e.g., Zug et al., 2001), ranging from tiny geckos to Komodo Dragons and Anacondas, and from fully limbed to limbless morphotypes (Evans, 2003), with specialized gliders, burrowers, climbers, runners and swimmers. Historically (Romer, 1956), squamates were divided into two groups, "Lacertilia" (lizards, Amphisbaenia) and Ophidia (now Serpentes, snakes), but phylogenetic analyses (e.g., Estes et al., 1988; Lee, 1998; Townsend et al., 2004; Vidal and Hedges, 2005) have shown that "Lacertilia" in this sense is not monophyletic. Use of the informal "lizard" is acceptable for a definitive squamate that is neither a snake nor an amphisbaenian, but Lacertilia should not be used.

The first cladistic analysis using morphological characters (Estes et al., 1988) divided Squamata into Iguania (pleurodont and acrodont lineages) and Scleroglossa (all non-iguanian squamates), and most morphological trees show a similar topology (e.g., Lee, 1998; Conrad and Norell, 2006; Sánchez-Martinez et al., 2007; Conrad, 2008). However, molecular trees (e.g., Townsend et al., 2004; Vidal and Hedges, 2005) nest Iguania within Scleroglossa (rendering the latter paraphyletic). Given the uncertainty in the relationships of "lizard" clades, Squamata is best defined as all lepidosaurs that are more closely related to snakes than to *Sphenodon*.

Timing the origins of major squamate groups (e.g., Iguania, Anguimorpha) depends on the tree used (morphological or molecular, Fig. 2.1) and the attribution of early taxa, notably *Tikiguana* (Carnian, India, Datta and Ray, 2006), *Bharatagama* and its pleurodont contemporary (Early Jurassic, India, Evans et al., 2002), and lizards from the Middle Jurassic of the UK (Evans, 1994, 1998), Central Asia (Nessov, 1988, Fedorov and Nessov, 1992; Martin et al., 2006), and China (Clark et al., 2006). Nonetheless, the first radiation must have occurred between the Late Triassic and Middle Jurassic. Many Jurassic-Early Cretaceous taxa are either stem-squamates or basal members of major clades, but mid-Cretaceous fossil squamates demonstrate increased morphological diversity (e.g., Evans et al., 2006; Li et al., 2007) and provide the first records of modern families (Evans, 2003). Whether this represents a real Cretaceous trend, or is simply a reflection of the more complete Cretaceous record remains to be determined.

2.3 The Lower Temporal Bar in Lepidosaurian Evolution

The fully diapsid skull and fixed quadrate of *Sphenodon* was long considered primitive by comparison with the open temporal region and streptostyly of squamates (e.g., Robinson, 1967b). However, a combination of new material (e.g., Evans, 1980; Whiteside, 1986) and new phylogenies (e.g., Gauthier et al., 1988; Müller, 2004) showed that the lower temporal bar was already absent in the last common ancestor

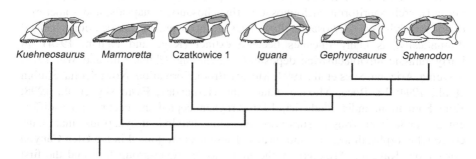

Kuehneosaurus   *Marmoretta*   *Czatkowice 1*   *Iguana*   *Gephyrosaurus*   *Sphenodon*

Fig. 2.3 Phylogenetic series of lepidosauromorph skulls in lateral view showing characters relating to the quadrate and lower temporal bar. Redrawn from Evans (2003), with the addition of the stem-lepidosaur (Czatkowice 1) from Poland (Evans and Borsuk-Białynicka, 2009)

of archosauromorphs and lepidosauromorphs (e.g., Müller, 2004), and more basal lineages, like the coelurosauravids (Evans, 1982; Evans and Haubold, 1987), also lack a bar (*Youngina* may have regained it). Thus the first lepidosauromorphs inherited a skull in which the ventral margin of the lower temporal fenestra was open, the quadratojugal was small, and the jugal lacked a posterior process (Fig. 2.3).

Enlargement of the quadrate conch occurred subsequently (as in *Paliguana*, kuehneosaurids, and the new Czatkowice genera). Squamates further modified the temporal region by reducing the bony links within the palatoquadrate (quadrate/epipterygoid), between the palatoquadrate and the rest of the skull (especially epipterygoid/pterygoid; quadrate/pterygoid joints), and between the braincase and the dermal skull roof – variously developing squamate metakinesis, streptostyly and mesokinesis, presumably as an aid to improved prey handling (e.g., Schwenk, 2000; Metzger, 2002). Rhynchocephalians followed a different trajectory, developing a more powerful, shearing bite (Robinson, 1976) and complex dentitions, and reacquiring a lower temporal bar as an adaptation to stabilize the quadrate (Whiteside, 1986; Fraser, 1988; Jones, 2006a, 2008; Moazen et al., 2009). Neither skull type is more primitive than the other, as demonstrated by the recent discovery of Late Cretaceous lizards from China with a complete lower temporal bar (Lü et al., 2008; Mo et al., 2009).

2.4 Discussion

2.4.1 Rhynchocephalians and Squamates

The contrast between the Triassic records of squamates and rhynchocephalians raises questions about taphonomy, habitat preferences, and palaeobiogeography (Evans, 1995, 2003). Rhynchocephalians occurred in both seasonally dry (e.g., the UK Triassic fissure assemblages, Fraser, 1985, 1994; Fraser and Walkden, 1983) and mesic habitats (e.g., Chinle Group, USA, Murry, 1987; Kaye and Padian, 1994; Heckert, 2004; Irmis, 2005), but typically in association with procolophonids, archosauriforms, trilophosaurs, rhynchosaurs, synapsids, and sometimes temnospondyls and phytosaurs. Post-Triassic, however, the association changed dramatically, as many lineages became extinct (Kaye and Padian, 1994). In Early Jurassic fluviolacustrine deposits (e.g., Lufeng, China, Luo and Wu, 1994; Kayenta, Arizona, Sues et al., 1994; McCoy Brook Formation, Nova Scotia, Shubin et al., 1994; La Boca, Mexico, Clark and Hernandez; Fastovsky et al., 1998; Kota Formation, India, Yadagiri, 1986), rhynchocephalians occur with a different assemblage of Triassic survivors, including tritylodont synapsids, mammals, crocodylomorphs, dinosaurs, and turtles. These taxa are joined in the Glen Canyon Formation (Kayenta, Arizona) by the first known crown-group frog and the first recorded gymnophionan (Sues et al., 1994), but no squamates, salamanders, or choristoderes, and in the Kota Formation, India, by very rare frogs and lizards (Yadagiri, 1986; Evans et al., 2002).

In the Carnian Tiki Formation of India, the only recorded Triassic lizard, *Tikiguana* (a single acrodont jaw), occurs in a typical Triassic assemblage like that described above (a phytosaur, a rauisuchid archosauriform, a rhynchosaur, and a non-mammalian cynodont); the only more derived taxon is an early mammal (Datta and Ray, 2006). However, apart from *Tikiguana*, *Bharatagama*, and fragmentary pleurodont remains from the Kota Formation, the earliest recorded lizards come from the Middle Jurassic of Laurasia (the UK, Kyrgyzstan, China, Evans, 1994, 1998; Fedorov and Nessov, 1992; Clark et al., 2006). From this time onwards, many squamates are found in mesic deposits with fish, amphibians (salamanders, albanerpetontids, frogs, rare caecilians), crocodiles, turtles, and frequently choristoderes. This stable assemblage persisted in Laurasia until the Miocene and typically represents lowland, freshwater lagoonal or wetland deposits (e.g., Middle Jurassic Forest Marble, UK, Evans and Milner, 1994; Late Jurassic, Guimarota, Portugal, Martin and Krebs, 2000; Early Cretaceous, Las Hoyas, Spain, Buscalioni and Fregenal-Martinez, 2006; Early Cretaceous Jehol Biota, China, Chang et al., 2003). No Triassic/Early Jurassic deposit has yielded an equivalent assemblage and thus the pre Middle Jurassic record of all component groups is poor. One key difference between the mesic Triassic/Early Jurassic Laurasian deposits that yield rhynchocephalians (e.g., Chinle Group, Glenn Canyon Formation), and those of the Middle Jurassic onward that produce squamates, is the presence in the latter but not the former of salamanders and choristoderes. These clades must have been present in the Triassic but are unrecorded (the choristoderan status of *Pachystropheus* [Storrs and Gower, 1993] is questionable). Rhynchocephalians and squamates are found together in some Jurassic/Cretaceous deposits, but rarely in equal proportions (Evans, 1995). Typically, where squamates are common, rhynchocephalians are rare (e.g., Kirtlington; Purbeck Limestone Group, UK) or absent (e.g., Guimarota), and vice versa (e.g., Solnhofen, Germany; Cerin, France). It is difficult to see how this distinction could be purely taphonomic and it may indicate a subtle difference between early squamate and rhynchocephalian ecology. Perhaps Triassic rhynchocephalians could tolerate a wider range of environmental conditions than early squamates, with this tolerance facilitating the post-Triassic survival of rhynchocephalians in more marginal habitats.

2.4.2 Evolution, Diversification and Extinction

Reisz and Müller (2004) proposed a molecular calibration date of 257–252 million years (Ma) for the archosauromorph-lepidosauromorph dichotomy, but this Late Permian date is difficult to reconcile with levels of Permo-Triassic archosauromorph diversity, or with the, admittedly poorer, lepidosauromorph record. More recent estimates placed the split ~299.8–259.7 Ma (Benton and Donoghue, 2007) or 303.9–263.0 Ma (Sanders and Lee, 2007), during the Lower-Middle Permian, and this is more congruent with the fossil evidence. Early lepidosauromorphs survived the end-Permian crisis, aided perhaps by small size and lower energy needs, and then radiated within the degraded Early Triassic ecosystems (Benton et al., 2004; Roopnarine et al., 2007). The well-nested phylogenetic position of the earliest known Carnian rhynchocephalians argues strongly for an unrecorded Middle Triassic history (Fig. 2.1) and a squamate – rhynchocephalian split in the Early to Middle Triassic (roughly consistent with hemoglobin analysis, Gorr et al., 1998). Rhynchocephalians apparently radiated first, achieving a global Late Triassic distribution (Jones et al., 2009). Both groups survived the end Triassic (or Carnian-Norian) extinctions and continued to diversify, but with differing fates. The Laurasian squamate record steadily improves throughout the Mesozoic. In contrast, rhynchocephalians are not recorded in Asia after the Early Jurassic, despite many apparently suitable small vertebrate localities (e.g., Nessov, 1988; Alifanov, 1993; Gao and Hou, 1996; Gao and Norell, 2000; Chang et al., 2003; Martin et al., 2006; Jones, 2006b). In Euramerica, they survived into the Early Cretaceous, but are unrecorded post-Albian. In the south, however, rhynchocephalians are known from the Late Cretaceous of South America (Apesteguía and Novas, 2003; Apesteguía, 2005a, b; Apesteguía and Rougier, 2007) and possibly the Palaeocene of Morocco (Augé and Rage, 2006). *Sphenodon* is recorded with certainty on New Zealand from the Pleistocene onwards (e.g., Crook, 1975; Holdaway and Worthy, 1997; Worthy, 1998; Worthy and Grant-Mackie, 2003), but rhynchocephalian jaw material is known from the Miocene (Jones et al., 2009). How long rhynchocephalians survived on other southern continents remains unknown (Apesteguía, 2005a).

Currently, the earliest known terrestrial snakes are from the Albian-Cenomanian of North America (Gardner and Cifelli, 1999); putative records from the Late Jurassic (Callison, 1987) are misidentified (SE, pers. obs.). However, in sharp contrast to lizards (e.g., Gao and Hou, 1996; Gao and Fox, 1996; Gao and Nessov, 1998; Gao and Norell, 2000), snakes are rare in the Mesozoic deposits of Euramerica and are unrecorded in Asia until well into the Cenozoic. By contrast, snakes are relatively common in southern continents from the Late Cretaceous onwards (e.g., Albino, 1996; Werner and Rage, 1994; Krause et al., 2003), and lizards are rare (e.g., Krause et al., 2003; Apesteguía et al., 2005; Apesteguía and Zaher, 2006). This has fuelled speculation that there were important differences between the Mesozoic lepidosaurian faunas of northern and southern continents (Apesteguía and Novas, 2003; Krause et al., 2003; Apesteguía, 2005a; Apesteguía and Zaher, 2006; Apesteguía and Rougier, 2007). The typical mesic lizard/salamander/choristodere assemblage that characterizes Laurasian microvertebrate horizons has not been recovered from Gondwana. That of Anoual (Early Cretaceous, Morocco, Sigogneau-Russell et al., 1998) is close, in that it contains fish, amphibians (frogs, caecilians, albanerpetontids), lizards, rhynchocephalians, turtles, crocodiles, dinosaurs, pterosaurs, and mammals, but it lacks both salamanders and choristoderes. Choristoderes have never been found in southern continents and, recent plethodontid range expansions excepted, the salamander record is limited to rare sirenid-like taxa from the Late Cretaceous of Sudan, Niger and South America (Evans et al., 1996). These differences between northern and southern small vertebrate assemblages are tantalizing, and warrant more detailed work.

Terrestrial squamates were little affected by the end-Cretaceous crisis, with the exception of a clade of large herbivorous Asian-American lizards (Evans, 2003), the boreoteiioids (*sensu* Nydam et al., 2007). The fate of the surviving Cretaceous Gondwanan rhynchocephalians is not yet known. *Sphenodon* eats a wide range of invertebrates and small vertebrates (e.g., Dawbin, 1962; Walls, 1981; Markwell, 1998; Cree et al., 1999; Moore and Godfrey, 2006), but known Late Cretaceous South American taxa (e.g., *Priosphenodon*, Apesteguia and Novas, 2003) were large herbivores. It is possible that the latter declined, like the herbivorous boreoteiioids, because of global cooling; digestion of plant material in reptiles being dependent on external ambient temperatures (Harlow et al., 1976; Troyer, 1987; van Marken Lichtenbelt, 1992; Tracy et al., 2005). The more generalist sphenodontines may have been better able to survive at the southern periphery.

2.5 Conclusions

Lepidosauromorphs probably diverged from archosauromorphs in the mid-Permian. They survived the end-Permian crisis and joined a depauperate Early Triassic fauna characterized by small, versatile tetrapods (the ancestors of lissamphibians, mammals, dinosaurs, etc.). Ancestral lepidosauromorphs had a skull without a lower temporal bar. Lepidosauria probably originated in the Early-Middle Triassic. Rhynchocephalians may have been the first lineage to achieve a global distribution, as part of a Late Triassic assemblage including procolophonians, phytosaurs, temnospondyls, synapsids, archosauromorphs and basal archosaurs. Unlike most of these groups, however, rhynchocephalians survived the end-Triassic (and/or Carnian/Norian) extinctions, but declined first in Asia (Early Jurassic) and then Euramerica (mid-Cretaceous) as limbed squamates diversified (although the two events are not necessarily causally linked, Jones, 2006b). In southern continents, rhynchocephalians survived into the Late Cretaceous (South America) and beyond, in association with a terrestrial assemblage that included abundant snakes but rarer lizards. Herbivorous rhynchocephalians (like herbivorous lizards) may have been more vulnerable to environmental changes (e.g., cold) at the Cretaceous-Palaeogene boundary than their opportunistic relatives, some of which survived and reached New Zealand (although it is not known when). Future discoveries in Mesozoic and Palaeogene deposits around the world will test these hypotheses but as Pamela Robinson recognized more than 50 years ago, India could be pivotal, having a unique palaeobiogeography (long isolation); important Mesozoic horizons, and tantalizing fossils (e.g., *Tikiguana, Bharatagama*).

Acknowledgments Our thanks to the conference organisers for the invitation to participate and to colleagues who have collaborated with us in work on fossil lepidosaurs, notably: Magdalena Borsuk-Białynicka (Poland); Dan Chure, David Krause (USA); Makoto Manabe (Japan); Alan Tennyson (New Zealand); Yuan Wang (China); and Trevor Worthy (Australia). Jerry D. Harris (USA) and Sebastian Apesteguía (Argentina) commented on an earlier draft of the manuscript. Although her phylogenetic interpretations have been revised, Pamela Robinson played a major role in focusing attention on lepidosauromorph evolution, and on the enormous potential of fissure infills and other microvertebrate assemblages in uncovering the history of the group.

References

Albino, AM (1996) The South American fossil Squamata (Reptilia: Lepidosauria). Münc Geowiss Abh, 30:185–202.

Alifanov, V (1993) Some peculiarities of the Cretaceous and Palaeogene lizard faunas of the Mongolian People's Republic. Kaupia, 3:9–13.

Apesteguía, S (2005a) Post-Jurassic sphenodontids: identity of the last lineages. In: Kellner, AWA, Henriques, DDR, Rodrigues, T (eds) Boletim de Resumos, II Congresso Latino-Americano de Paleontologia de Vertebrados. Museu Nacional, Rio de Janiero, Brazil.

Apesteguía, S (2005b) A Late Campanian sphenodontid (Reptilia, Diapsida) from northern Patagonia. C R Palevol, 4:663–669.

Apesteguía, S, Agnolin, FC, Lio, GL (2005) An early Late Cretaceous lizard from Patagonia, Argentina. C R Palevol, 4:311–315.

Apesteguía, S, Novas, FE (2003) Large Cretaceous sphenodontian from Patagonia provides insight into lepidosaur evolution in Gondwana. Nature, 425:609–612.

Apesteguía, S, Rougier, GW (2007) A Late Campanian sphenodontid maxilla from Northern Patagonia. Am Mus Novit, 3581:1–11.

Apesteguía, S, Zaher, H (2006) A Cretaceous terrestrial limbed snake with robust hindlimbs and sacrum. Nature, 440:1037–1040.

Augé, M, Rage, J-C (2006) Herpetofauna from the upper Paleocene and lower Eocene of Morocco. Ann Paléontol, 92:235–253.

Bannock, CA, Whitaker, AH, Hickling, GJ (1999) Extreme longevity of the common gecko (*Hoplodactylus maculatus*) on Motunau Island, Canterbury, New Zealand. N Z J Ecol, 23:101–103.

Bartholomai, A (1979) New lizard-like reptiles from the Early Triassic of Queensland. Alcheringa, 3:225–234.

Benton, MJ (1985) Classification and phylogeny of the diapsid reptiles. Zool J Linn Soc, 84:97–164.

Benton, MJ, Donoghue, PCJ (2007) Paleontological evidence to date the tree of life. Mol Biol Evol, 24:26–53.

Benton, MJ, Tverdokhlebov, VP, Surkov, MV (2004) Ecosystem remodelling among vertebrates at the Permo-Triassic boundary in Russia. Nature, 432:97–100.

Borsuk-Białynicka, M, Cook, E, Evans, SE, Maryanska, T (1999) A microvertebrate assemblage from the Early Triassic of Poland. Acta Palaeontol Polonica, 44:167–188.

Buscalioni, AD, Fregenal-Martinez, M (2006) Archosaurian size bias in Jurassic and Cretaceous freshwater ecosystems. In: Barrett, PM, Evans, SE (eds) Proceedings of the 9th international symposium on Mesozoic terrestrial ecosystems and biota, abstract and proceeding. Natural History Museum, London, UK.

Callison, G (1987) Fruita: a place for wee fossils. In: Averett, WR (ed) Paleontology and geology of the Dinosaur Triangle: guidebook for 1987 field trip. Museum of Western Colorado, Grand Junction, CO, pp. 91–96.

Carroll, RL (1975) Permo-Triassic 'lizards' from the Karroo. Palaeontol Afr, 18:71–87.

Carroll, RL (1982) A short-limbed lizard from the *Lystrosaurus* zone (Lower Triassic) of South Africa. J Paleontol, 56:183–190.

Carroll, RL, Wild, R (1994) Marine members of the Sphenodontia. In: Fraser, NC, Sues, H-D (eds) In the shadow of the dinosaurs: early Mesozoic tetrapods. Cambridge University Press, Cambridge, MA.

Chang, MM, Chen, PJ, Wang, YQ, Wang, Y (2003) The Jehol Biota: emergence of feathered dinosaurs and beaked birds. Shanghai Scientific and Technical Publishers, Shanghai, China.

Clark, JM, Hernandez, RR (1994) A new burrowing diapsid from the Jurassic La Boca Formation of Tamaulipas, Mexico. J Vertebr Paleontol, 14:180–195.

Clark, JM, Xing, X, Eberth, DA, Forster, CA, Malkus, M, Hemming, S, Hernandez, R (2006) The Middle-Late Jurassic terrestrial transition: new discoveries from the Shishugou Formation,

Xinjiang, China. In: Barrett, PM, Evans, SE (eds) Proceedings of the 9th international symposium on Mesozoic terrestrial ecosystems and biota. Natural History Museum, London, UK.

Cocude-Michel, M (1963) Les rhynchocephales et les sauriens des calcaires lithographiques (Jurassique superieur) d'Europe Occidentale. Nouv Arch Mus Hist Nat Lyon, 7:1–187.

Colbert, EH (1966) *Icarosaurus*, a gliding reptile from the Triassic of New Jersey. Am Mus Novit, 2246:1–23.

Colbert, EH (1970) The Triassic gliding reptile *Icarosaurus*. Bull Am Mus Nat Hist, 143:85–142.

Colbert, EH, Olsen, PE (2001) A new and unusual aquatic reptile from the Lockatong Formation of New Jersey (Late Triassic, Newark Supergroup). Am Mus Novit, 3334:1–24.

Conrad, J (2008) Phylogeny and systematics of Squamata (Reptilia) based on morphology. Bull Am Mus Nat Hist, 310:1–182.

Conrad, JL, Norell, MA (2006) High-resolution X-ray computed tomography of an Early Cretaceous gekkonomorph (Squamata) from Öösh (Ovorkhangai; Mongolia). Hist Biol, 18:405–431.

Cree, A (1994) Low annual reproductive output in female reptiles from New Zealand. N Z J Zool, 21:351–372.

Cree, A, Lyon, G, Cartland Shaw, L, Tyrrel, C (1999) Stable isotope ratios as indicators of marine versus terrestrial inputs to the diets of wild and captive tuatara (*Sphenodon punctatus*). N Z J Zool, 26:243–253.

Crook, IG (1975) The tuatara. In: Kuschel, G (ed) Biogeography and ecology in New Zealand. Junk, Hague, The Netherlands, pp. 331–352.

Datta, PM, Ray, S (2006) Earliest lizard from the Late Triassic (Carnian) of India. J Vertebr Paleontol, 26:795–800.

Dawbin, WH (1962) The tuatara in its natural habitat. Endeavour, 81:16–24.

de Braga, M, Rieppel, O (1997) Reptile phylogeny and the relationships of turtles. Zool J Linn Soc, 120:281–354.

Dilkes, DW (1998) The Early Triassic rhynchosaur *Mesosuchus browni* and the interrelationships of basal archosauromorph reptiles. Phil Trans R Soc, B, 353:501–541.

Dzik, J, Sulej, T (2007) A review of the early Late Triassic Krasiejow biota from Silesia, Poland. Acta Palaeontol Polonica, 64:3–27.

Estes, R, De Queiroz, K, Gauthier, J (1988) Phylogenetic relationships within Squamata. In: Estes, R, Pregill, G (eds) Phylogenetic relationships of the lizard families. Stanford University Press, Stanford, CA.

Evans, SE (1980) The skull of a new eosuchian reptile from the Lower Jurassic of South Wales. Zool J Linn Soc, 70:203–264.

Evans, SE (1982) Gliding reptiles of the Upper Permian. Zool J Linn Soc, 76:97–123.

Evans, SE (1984) The classification of the Lepidosauria. Zool J Linn Soc, 82:87–100.

Evans, SE (1988) The early history and relationships of the Diapsida. In: Benton, MJ (ed) The phylogeny and classification of the tetrapods. Oxford University Press, Oxford.

Evans, SE (1991) A new lizard-like reptile (Diapsida: Lepidosauromorpha) from the Middle Jurassic of Oxfordshire. Zool J Linn Soc, 103:391–412.

Evans, SE (1994) A new anguimorph lizard from the Jurassic and Lower Cretaceous of England. Palaeontol, 37:33–49.

Evans, SE (1995) Lizards: evolution, early radiation and biogeography. In: Sun, A, Wang, Y (eds) Proceedings of the 6th symposium on Mesozoic terrestrial ecosystems and Biota. Short pap, Ocean Press, Beijing, China.

Evans, SE (1998) Crown group lizards from the Middle Jurassic of Britain. Palaeontogr, A250:123–154.

Evans, SE (2001) The Early Triassic 'lizard' *Colubrifer campi*: a reassessment. Palaeontol, 44:1033–1041.

Evans, SE (2003) At the feet of the dinosaurs: the early history and radiation of lizards. Biol Rev, 78:513–551.

Evans, SE (2009) An early kuehneosaurid reptile (Reptilia: Diapsida) from the Early Triassic of Poland. Palaeontol Polonica, 65:145–178.

Evans, SE, Borsuk-Białynicka, M (2009) A small lepidosauromorph reptile from the early Triassic of Poland. Palaeontol Polonica, 65:179–202.

Evans, SE, Haubold, H (1987) A review of the Upper Permian genera *Coelurosauravus, Weigeltisaurus* and *Gracilisaurus* (Reptilia: Diapsida). Zool J Linn Soc, 90:275–303.

Evans, SE, Manabe, M, Noro, M, Isaji, S, Yamaguchi, M (2006) A long-bodied aquatic varanoid lizard from the Early Cretaceous of Japan. Palaeontol, 49:1143–1165.

Evans, SE, Milner, AR (1994) Microvertebrate faunas from the Middle Jurassic of Britain. In: Fraser, NC, Sues, H-D (eds) In the shadow of the dinosaurs: early Mesozoic tetrapods. Cambridge University Press, Cambridge, MA.

Evans, SE, Milner, AR, Werner, C (1996) Sirenid salamanders and a gymnophionan from the Late Cretaceous of the Sudan. Palaeontol, 39:77–95.

Evans, SE, Prasad, GVR, Manhas, BK (2002) Fossil lizards from the Jurassic Kota Formation of India. J Vertebr Paleontol, 22:299–312.

Fastovsky, DE, Bowring, SA, Hermes, OD (1998) Radiometric age dates for the La Boca vertebrate assemblage (late Early Jurassic), Huizachal Canyon, Tamaulipas, Mexico: Avances en Investigación: Paleontología de Vertebrados, Inst Invest Ciencias Tierra. Universidad Autónoma del Estado de Hidalgo Spec Publ, 1:4–11.

Fedorov, PV, Nessov, LA (1992) A lizard from the boundary of the Middle and Late Jurassic of north-east Fergana. Bull St. Petersburg Univ, Geol Geogr, 3:9–14 [In Russian].

Ferigolo, J (1999) South American first record of a sphenodontian (Lepidosauria, Rhynchocephalia) from Late Triassic–Early Jurassic of Rio Grande do Sul State, Brazil. In: Leanza, HA (ed) Proceedings of the 7th International Symposium on Mesozoic Terrestrial Ecosystem (abstract). Museo Argentino de Ciencias Naturales, Buenos Aires, Argentina.

Flynn, JJ, Fox, SR, Parrish, JM, Ranivoharimanana, L, Wyss, AR (2006) Assessing diversity and paleoecology of a Middle Jurassic microvertebrate assemblage from Madagascar. Natl Mus Nat Hist Sci Bull, 37:476–489.

Fraser, NC (1982) A new rhynchocephalian from the British Upper Triassic. Palaeontol, 25: 709–725.

Fraser, NC (1985) Vertebrate faunas from Mesozoic fissure deposits of South West Britain. Mod Geol, 9:273–300.

Fraser, NC (1988) The osteology and relationships of *Clevosaurus* (Reptilia, Sphenodontida). Phil Trans R Soc Lond, B312:125–178.

Fraser, NC (1994) Assemblages of small tetrapods from British Late Triassic fissure deposits. In: Fraser, NC, Sues, H-D (eds) In the shadow of the dinosaurs: early Mesozoic tetrapods. Cambridge University Press, Cambridge, MA.

Fraser, NC, Benton, MJ (1989) The Triassic reptiles *Brachyrhinodon* and *Polysphenodon* and the relationships of the sphenodontids. Zool J Linn Soc, 96:413–445.

Fraser, NC, Walkden, GM (1983) The ecology of a Late Triassic reptile assemblage from Glouchestershire, England. Palaeogeogr Palaeoclimatol Palaeoecol, 42:341–365.

Gans, C (1983) Is *Sphenodon punctatus* a maladapted relic? In: Rhodin, AGJ, Miyata, K (eds) Advances in herpetology and evolutionary biology. Harvard University Press, Cambridge, MA.

Gao, KQ, Fox, RC (1996) Taxonomy and evolution of Late Cretaceous lizards (Reptilia: Squamata) from western Canada. Bull Carnegie Mus Nat Hist, 33:1–107.

Gao, KQ, Hou, LH (1996) Systematics and diversity of squamates from the Upper Cretaceous Djadochta Formation, Bayan Mandahu, Gobi Desert, People's Republic of China. Can J Earth Sci, 33:578–598.

Gao, KQ, Nessov, LA (1998) Early Cretaceous squamates from the Kyzylkum Desert, Uzbekistan. N J Geol Paläontol Abh, 207:289–309.

Gao, KQ, Norell, MA (2000) Taxonomic composition and systematics of Late Cretaceous lizard assemblages from Ukhaa Tolgod and adjacent localities, Mongolian Gobi Desert. Bull Am Mus Nat Hist, 249:1–118.

Gardner, JD, Cifelli, RL (1999) A primitive snake from the Cretaceous of Utah. Sp Pap Palaeontol, 60:87–100.

Gauthier, J, Estes, R, de Queiroz, K (1988) A phylogenetic analysis of the Lepidosauromorpha. In: Estes, R and Pregill, G (eds) Phylogenetic relationships of the lizard families. Stanford University Press, Stanford, CA.

Gorr, TA, Mable, BK, Kleinschmidt, T (1998) Phylogenetic analysis of the reptile haemoglobins: trees, rates and divergences. J Mol Evol, 47:471–485.

Gradstein, FW, Ogg, JG (2004) Geologic timescale 2004 – why, how and where next. Lethaia, 37:175–181.

Günther, A (1867) Contribution to the anatomy of *Hatteria* (*Rhynchocephalus*, Owen). Phil Trans R Soc Lond, 157:1–34.

Harlow, HJ, Hillman, SS, Hoffman, M (1976) The effect of temperature on digestive efficiency in the herbivorous lizard, *Dipsosaurus dorsalis*. J Comp Physiol B: Biochem, Syst Environ Physiol, 111:1–6.

Heckert, AB (2004) Late Triassic microvertebrates from the lower Chinle Group (Otischalkian-Adamanian: Carnian) southwestern USA. Natl Mus Nat Hist Sci Bull, 27: 1–170.

Hedges, SB, Poling, LL (1999) A molecular phylogeny of reptiles. Science, 283:998–1001.

Hill, RV (2005) Integration of morphological data sets for phylogenetic analysis of Amniota: the importance of integumentary characters and increased taxonomic sampling. Syst Biol, 54:530–547.

Holdaway, RN, Worthy, TH (1997) A reappraisal of the late Quaternary fossil vertebrates of Pyramid Valley Swamp, North Canterbury, New Zealand. N Z J Zool, 24:69–121.

Irmis, RB (2005) The vertebrate fauna of the upper Triassic Chinle formation in northern Arizona. Mesa Southwest Mus Bull, 6:63–88.

Jones, MEH (2004) Exploring rhynchocephalian skull morphology with morphometrics. J Vertebr Palaeontol, 24:76–77.

Jones, MEH (2006a) Skull evolution and functional morphology of *Sphenodon* and other Rhynchocephalia (Diapsida, Lepidosauria). Unpublished Ph.D. thesis, University of London, London, UK.

Jones, MEH (2006b) The Early Jurassic clevosaurs from China (Diapsida: Lepidosauria). Natl Mus Nat Hist Sci Bull, 37:548–562.

Jones, MEH (2006c) Tooth diversity and function in the Rhynchocephalia (Diapsida: Lepidosauria). In: Barrett, PM, Evans, SE (eds) Proceedings of the 9th International Symposium on Mesozoic Terrestrial Ecosystem and Biota. Natural History Museum, London, UK.

Jones, MEH (2008) Skull shape and feeding strategy in *Sphenodon* and other Rhynchocephalia (Diapsida: Lepidosauria). J Morphol, 269:945–966.

Jones, MEH, Tennyson, AJD, Evans, SE, Worthy, TH (2009) A sphenodontine (Rhynchocephalia) from the Miocene of New Zealand and palaeobiogeography of the tuatara (*Sphenodon*). Proc R Soc, B276:1385–1390.

Kaye, FT, Padian, K (1994) Microvertebrates from the Placerias Quarry: a window on the Late Triassic vertebrate diversity in the American South West. In: Fraser, NC, Sues, H-D (eds) In the Shadow of the Dinosaurs: early Mesozoic tetrapods. Cambridge University Press, Cambridge, MA.

Krause, D, Evans, SE, Gao, K (2003) First definitive record of a Mesozoic lizard from Madagascar. J Vertebr Paleontol, 23:842–856.

Laurin, M (1991) The osteology of a Lower Permian eosuchian from Texas and a review of diapsid phylogeny. Zool J Linn Soc, 101:59–95.

Lee, MSY (1998) Convergent evolution and character correlation in burrowing reptiles: towards a resolution of squamate phylogeny. Biol J Linn Soc, 65:369–453.

Li, PP, Gao, KQ, Hou, LH, Xu, X (2007) A gliding lizard from the Early Cretaceous of China. Proc Natl Acad Sci, 104:5507–5509.

Lü, J-C, Ji, S-A, Dong, Z-M, Wu, X-C (2008) An Upper Cretaceous lizard with a lower temporal arcade. Naturewissenschaften, 95:663–669.

Luo, Z, Wu, X-C (1994) The small tetrapods of the Lower Lufeng Formation, Yunnan, China. In: Fraser, NC, Sues, H-D (eds) In the Shadow of the Dinosaurs: early Mesozoic tetrapods. Cambridge University Press, Cambridge, MA.

Markwell, TJ (1998) Relationship between tuatara *Sphenodon punctatus* and fairy prion *Pachyptila turtur* densities in different habitats on Takapourewa (Stephens Island), Cook Strait, New Zealand. Mar Ornithol, 26:81–83.

Martin, T, Averianov, AO, Pfretzschner, H-U (2006) Palaeoecology of the Middle to Late Jurassic vertebrate assemblages from the Fergana and Jungar Basins (Central Asia). In: Barrett, PM, Evans, SE (eds) Proceedings of the 9th International Symposium on Mesozoic Terrestrial Ecosystems and Biota, Abstract & Proceeding. Natural History Museum, London, UK.

Martin, T, Krebs, B (2000) Guimarota, a Jurassic ecosystem. Dr Friedrich Pfeil, München, Germany, 155 pp.

Metzger, K (2002) Cranial kinesis in lepidosaurs: skulls in motion. In: Aerts, P, D'Août, K, Herrel, A, van Damme, R (eds) Topics in functional and ecological vertebrate morphology. Shaker publishing, Maastricht, The Netherlands.

Mo, J, Xu, X, Evans, SE (2009) The evolution of the lepidosaurian lower temporal bar: new perspectives from the Late Cretaceous of South China. Proc R Soc B, Spec Vol. Chinese Fossils, 277:331–336. (doi: 10.1098/rspb.2009.0030).

Moazen, M, Curtis, N, O'Higgins, P, Evans, SE, Fagan, M (2009) The function of the lower temporal bar in lepidosaurian evolution. Proc Natl Acad Sci, 20:8273–8277.

Modesto, SP, Reisz, RR (2002) An enigmatic new diapsid reptile from the Upper Permian of Eastern Europe. J Vertebr Paleontol, 22:851–855.

Moore, JA, Godfrey, SS (2006) *Sphenodon punctatus* (common tuatara). Opportunistic predation. Herpetol Rev, 37:81–82.

Müller, J (2004) The relationships among diapsid reptiles and the influence of taxon selection. In: Arratia, G, Wilson, MVH, Cloutier, R (eds) Recent advances in the origin and early radiation of vertebrates. Dr Friedrich Pfeil, München, Germany.

Murry, PA (1987) New reptiles from the Upper Triassic Chinle Formation of Arizona. J Paleontol, 61:773–786.

Nessov, LA (1988) Late Mesozoic amphibians and lizards of Soviet Middle Asia. Acta Zool Cracoviensia, 31:475–486.

Nydam, RL, Eaton, JG, Sankey, J (2007) New taxa of transversely-toothed lizards (Squamata: Scincomorpha) and new information on the evolutionary history of 'teiids'. J Paleontol, 81:538–549.

Parrington, FR (1935) On *Prolacerta broomi* gen. et sp. n., and the origin of lizards. Ann Mag Nat Hist, 16:197–205.

Reisz, RR, Müller, J (2004) Molecular timescales and the fossil record: a paleontological perspective. Trends Genet, 20:237–241.

Renesto, S, Posenato, R (2003) A new lepidosauromorph reptile from the Middle Triassic of the Dolomites (Northern Italy). Riv Ital Paleont Strat, 109:463–474.

Rest, JS, Ast, JC, Austin, CC, Waddell, PJ, Tibbettes, EA, Hay, JM, Mindell, DP (2003) Molecular systematics of primary reptilian lineages and the tuatara mitochondrial genome. Mol Phylogeny Evol, 29:289–297.

Reynoso, VH (1996) A Middle Jurassic *Sphenodon*-like sphenodontian (Diapsida: Lepidosauria) from Huizachal Canyon, Tamaulipas, Mexico. J Vertebr Paleontol, 16:210–221.

Reynoso, VH (1997) A 'beaded' sphenodontian (Diapsida: Lepidosauria) from the Early Cretaceous of central Mexico. J Vertebr Paleontol, 17:52–59.

Reynoso, VH (2000) An unusual aquatic sphenodontian (Reptilia: Diapsida) from the Tlayua Formation (Albian), Central México. J Vertebr Paleontol, 74:133–148.

Reynoso, VH (2005) Possible evidence of a venom apparatus in a Middle Jurassic sphenodontian from the Huizachal red beds of Tamaulipas, México. J Vertebr Paleontol, 25:646–654.

Reynoso, VH, Clark, JM (1998) A dwarf sphenodontian from the Jurassic La Boca Formation of Tamaulipas, Mexico. J Vertebr Paleontol, 18:333–339.

Robinson, PL (1962) Gliding lizards from the Upper Keuper of Great Britain. Proc Geol Soc Lond, 1601:137–146.

Robinson, PL (1967a) Triassic vertebrates from upland and lowland. Sci Cult, 33:169–173.

Robinson, PL (1967b) The evolution of the Lacertilia. Coll Int CNRS, 163:395–407.

Robinson, PL (1976) How *Sphenodon* and *Uromastix* grow their teeth and use them. In: Bellairs, Ad'A, Cox, CB (eds) Morphology and biology of the reptiles. Academic Press, London, UK.

Romer, AS (1956) Osteology of the reptiles. University of Chicago Press, Chicago, IL.

Roopnarine, PD, Angielczyk, KD, Wang, SC, Hertog, R (2007) Trophic network models explain instability of Early Triassic terrestrial communities. Proc R Soc, B274:2077–2086.

Russmusen, TE, Callison, G (1981) A new herbivorous sphenodontid (Rhynchocephalia: Reptilia) from the Jurassic of Colorado. J Paleontol, 55:1109–1116.

Sanchez-Martinez, PM, Ramirez-Pinilla, MP, Miranda-Esquivel, DR (2007) Comparative histology of the vaginal-cloacal region in Squamata and its phylogenetic implications. Acta Zool, 88:289–307.

Sanders, KL, Lee, MSY (2007) Evaluating molecular clock calibrations using Bayesian analyses with soft and hard bounds. Biol Lett, 3:275–279.

Schwenk, K (2000) Feeding: form, function, and evolution in tetrapod vertebrates. Academic Press, San Diego, CA.

Shubin, NH, Olsen, PE, Sues, H-D (1994) Early Jurassic small tetrapods from the McCoy Brook Formation of Nova Scotia, Canada. In: Fraser, NC, Sues, H-D (eds) In the shadow of the dinosaurs: early Mesozoic tetrapods. Cambridge University Press, Cambridge, MA.

Sigogneau-Russell, D, Evans, SE, Levine, J, Russell, D (1998) An Early Cretaceous small vertebrate assemblage from the Early Cretaceous of Morocco. N M Mus Nat Hist Sci, 14:177–182.

Smith, R, Botha, J (2005) The recovery of terrestrial vertebrate diversity in the South African Karoo Basin after the end-Permian extinction. C R Palevol, 4:623–636.

Spencer, PS, Storrs, GW (2002) A re-evaluation of small tetrapods from the Middle Triassic Otter Sandstone Formation of Devon, England. Palaeontol, 45:447–467.

Storrs, GW, Gower, DJ (1993) The earliest possible choristodere (Diapsida) and gaps in the fossil record of semi-aquatic reptiles. J Geol Soc, 150:1103–1107.

Sues, H-D, Baird, D (1993) A skull of a sphenodontian lepidosaur from the New Haven Arkose (Upper Triassic, Norian) of Connecticut. J Vertebr Paleontol, 13:370–372.

Sues, H-D, Clark, JM, Jenkins, FA, Jr (1994) A review of the Early Jurassic tetrapods from the Glen Canyon Group of the American Southwest. In: Fraser, NC, Sues, H-D (eds) In the shadow of the dinosaurs: early Mesozoic tetrapods. Cambridge University Press, Cambridge, MA.

Sun, AL, Li, JL, Ye, XK, Dong, ZM, Hou, LH (1992) The Chinese fossil reptiles and their kins. Science Press, Beijing, China.

Tatarinov, LP (1978) Triassic prolacertilians of the USSR. Paleontol J, 4:505–514.

Thompson, MB, Daugherty, CH (1998) Metabolism of tuatara, *Sphenodon punctatus*. Comp Biochem Physiol, 119A:519–522.

Throckmorton, GS, Hopson, JA, Parks, P (1981) A redescription of *Toxolophosaurus cloudi* Olson, a Lower Cretaceous herbivorous sphenodontian reptile. J Paleontol, 55:586–597.

Townsend, TM, Larson, A, Louis, E, Macey, JR (2004) Molecular phylogenetics of Squamata: the position of snakes, amphisbaenians, and dibamids, and the root of the squamate tree. Syst Biol, 53:735–757.

Tracy, CR, Flack, KM, Zimmerman, LC, Espinoza, RE, Tracy, CR (2005) Herbivory imposes constraints on voluntary hypothermia in lizards. Copeia, 2005:12–19.

Troyer, K (1987) Small differences in daytime body temperature affect digestion of natural food in a herbivorous lizard (*Iguana iguana*). Comp Biochem Physiol, 87:623–626.

van Marken Lichtenbelt, WD (1992) Digestion in an ectothermic herbivore, the green iguana (*Iguana iguana*) – effect of food composition and body temperature. Physiol Zool, 65:649–673.

Vidal, N, Hedges, SB (2005) The phylogeny of squamate reptiles (lizards, snakes, and amphisbaenians) inferred from nine nuclear protein-coding genes. C R Biol, 328:1000–1008.

Walls, GY (1981) Feeding ecology of the tuatara (*Sphenodon punctatus*) on Stephen Island, Cook Strait. N Z J Ecol, 4:89–97.

Watson, DMS (1914) *Pleurosaurus* and the homologies of the bones of the temporal region of the lizard's skull. Ann Mag Nat Hist, 14:84–95.

Werner, C, Rage, J-C (1994) Mid-Cretaceous snakes from Sudan. A preliminary report on an unexpectedly diverse snake fauna. C R Acad Sci, 319:247–252.

Whiteside, DI (1986) The head skeleton of the Rhaetian sphenodontid *Diphydontosaurus avonis* gen. et sp. nov., and the modernising of a living fossil. Phil Trans R Soc Lond, B312:379–430.

Wilkinson, M, Benton, MJ (1996) Sphenodontid phylogeny and the problems of multiple trees. Phil Trans R Soc Lond, 351:1–16.

Worthy, TH (1998) Quaternary fossil faunas of Otago, South Island, New Zealand. J R Soc N Z, 28:421–521.

Worthy, TH, Grant-Mackie, JA (2003) Late-Pleistocene avifaunas from Cape Wanbrow, Otago, South Island, New Zealand. J R Soc N Z, 33:427–485.

Wu, X-C (2003) Functional morphology of the temporal region in the Rhynchocephalia. Can J Earth Sci, 40:589–607.

Yadagiri, P (1986) Lower Jurassic lower vertebrates from Kota Formation, Pranhita-Godavari Valley, India. J Palaeontol Soc India, 31:89–96.

Zug, GR, Vitt, LJ, Caldwell, JP (2001) Herpetology, 2nd edn. Academic Press, San Diego, CA.

Chapter 3
Rahiolisaurus gujaratensis, n. gen. n. sp., A New Abelisaurid Theropod from the Late Cretaceous of India

Fernando E. Novas, Sankar Chatterjee, Dhiraj K. Rudra, and
P.M. Datta

3.1 Introduction

Abelisaurids are probably the most distinctive predatory dinosaurs from disjunct Gondwanan landmasses during the Cretaceous period. As far as the Indian record is concerned, abelisauroid remains have been discovered sporadically for the last 75 years from the Upper Cretaceous Lameta Formation of central and western India, immediately below the Deccan Trap lava flows. Indian abelisauroid taxa include several species that are based on fragmentary remains including *Indosuchus raptorius*, *Indosaurus matleyi, Lametasaurus indicus, Laevisuchus indicus*, and *Rajasaurus narmadensis* (Huene and Matley, 1933; Chatterjee, 1978; Wilson et al., 2003), but their anatomy and relationships are beginning to emerge with the description of new material and review of previous collections (Wilson et al., 2003; Novas and Bandyopadhyay, 1999, 2001; Novas et al., 2004; Carrano and Sampson, 2008).

The Lameta Formation is a fluvio-lacustrine coastal plain deposit about 50 m thick, and is well known for its dinosaur fauna. In 1996, Chatterjee and Rudra reported the discovery of a dinosaur graveyard in a mudstone facies of the Lameta Formation near Rahioli village, Kheda District, Gujarat. This bone bed has yielded assorted but disarticulated dinosaur bones of abelisaurids and titanosaurs. The age of the Lameta Formation is regarded as Maastrichtian on the basis of microfossils, vertebrates, and the associated basal flows of the Deccan lavas (Sahni and Bajpai, 1991; Chatterjee and Rudra, 1996).

The partial skeleton of a sympatric abelisaurid species *Rajasaurus narmadensis* was collected from a nearby area at the Temple Hill locality (Wilson et al., 2003).

The purpose of this paper is to describe this new material of *Rahiolisarus* and discuss its affinity with other Gondwana abelisauroids.

F.E. Novas (✉)
Conicet, Museo Argentino de Ciencias Naturales, Buenos Aires 1405, Argentina
e-mail: fernovas@yahoo.com.ar

S. Bandyopadhyay (ed.), *New Aspects of Mesozoic Biodiversity*,
Lecture Notes in Earth Sciences 132, DOI 10.1007/978-3-642-10311-7_3,
© Springer-Verlag Berlin Heidelberg 2010

3.2 Materials and Methods

A large number of abelisaurid bones were excavated during 1995 and 1997 expeditions from a single quarry of 50 m^2. The collection includes cervical, dorsal, sacral and caudal vertebrae, portions of pectoral and pelvic girdles, and several hind limb bones. Documentation of seven different-sized right tibiae suggests that the assemblage was formed by at least seven individuals at different ontogenetic stages. In spite of size gradation representing growth series, several duplicate bones in the collection (e.g. ilia, pubes, femora and tibiae) exhibit similar morphological features of typical abelisauroid traits, but hardly any taxonomic variation; we interpret that the entire theropod collection from this quarry may be referred to a single species. Individual bones of the new abelisaurid are given separate catalogue numbers. A partial association of pelvic elements and a femur in the field is represented by a right ilium (ISIR 550), a right pubis (554), and a right femur (557). Similarly, an axis (ISIR 658) was found in articulation with cervicals 3 (ISIR 659) and 4 (ISIR 660).

Rahiolisaurus possesses a number of synapomorphies that support its membership in Abelisauroidea (e.g. Carrano et al., 2002; Carrano and Sampson, 2008; Wilson et al., 2003; Sereno et al., 2004). These include an axial neural arch triangular and transversely expanded in dorsal aspect; postaxial cervical vertebrae with a prominent ridge connecting two zygapophyses; pubis and ilium tightly fused; ilium relatively low with postacetabular process bearing a sharp caudodorsal prominence; pubis transversely narrow and bearing a distal foot; and ischia massive and partially fused at the symphysis.

Institutional Abbreviations. ISI, Indian Statistical Institute, Kolkata; MACN-CH, Museo Argentino de Ciencias Naturales Bernardino Rivadavia, Paleontología de Vertebrados (Colección Chubut), Buenos Aires; MUCPv, Museo Universidad Nacional del Comahue, Paleovertebrados, Neuquén; PVL, Instituto Miguel Lillo, Tucumán.

3.3 Systematic Palaeontology

THEROPODA Marsh, 1881
CERATOSAURIA Marsh, 1884
ABELISAUROIDEA Bonaparte, 1991
ABELISAURIDAE Bonaparte and Novas, 1985
RAHIOLISAURUS, new genus
Rahiolisaurus gujaratensis, new genus and species

Etymology Named after the village of Rahioli, the fossil site; the specific epithet refers (Hindi) to the denizens of Gujarat province.

Holotype ISIR 550 (a right ilium), ISIR 554 (a right pubis), and ISIR 557 (a right femur) that were found in association. These bones are at the collection of the Geology Museum, Indian Statistical Institute, Kolkata.

Referred specimens ISIR 401–433, 435–454 collected in 1995 fieldwork, and ISIR 464, 465, 474, 475, 486–553, 555–556, 558–602, 645, 649, 657–660 in 1997 fieldwork, housed at the collection of the Geology Museum, Indian Statistical Institute, Kolkata.

Type locality Near Rahioli village (23° 3′ 26.2″ N, 73° 20′ 30.8″ E), Kheda District, Gujarat, western India.

Horizon Mudstone unit of the Lameta Formation, Late Cretaceous (Maastrichtian).

Diagnosis A slender-limbed abelisaurid theropod with: (1) premaxillary interdental plates fused and lacking vertical ridges; (2) dental foramina absent; (3) premaxillary teeth with teardrop-shaped cross section; a faint mesial keel but a rounded distal edge; (4) iliac blade with deep caudal notch on postacetabular process; (5) metatarsal I rod-like; and (6) metatarsal II strongly narrow proximally.

3.4 Description of *Rahiolisaurus*

3.4.1 Premaxilla and Teeth

Among skull elements, only the right premaxilla (ISIR 401) is preserved. The alveolar component bears four partial teeth. The premaxilla has a subrectangular base with nearly vertical rostral and caudal margins but the ascending nasal process is broken away (Fig. 3.1). The preserved base is a thick bone, which is gently convex laterally reflecting the rounding process of the snout. The lateral surface exhibits a pattern of external rugosities that is usually present in other

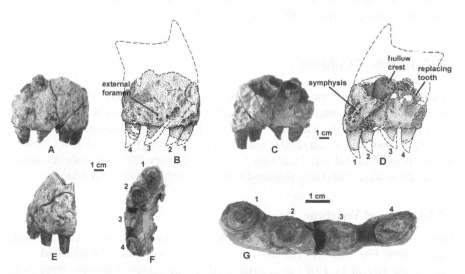

Fig. 3.1 Right premaxilla and associated teeth (ISIR 401) of *Rahiolisaurus gujaratensis*. (**a** and **b**) Lateral, (**c** and **d**), medial, (**e**), rostral, and (**f**), ventral views. (**g**) Cross-section of premaxillary teeth as seen in occlusal view

abelisaurids (e.g. *Abelisaurus*, *Carnotaurus*, *Majungasaurus*) and contains a row of small neurovascular foramina just above the alveolar margin.

The medial surface (apparently keeping most of its pristine condition) is smooth and lacks ornamentation on the interdental plates; in all other abelisaurids in which this region is preserved, the interdental plates are adorned with a distinctive pattern of vertical ridges. Most likely, interdental plates are fused in *Rahiolisaurus* to form a single bone on the lingual surface of premaxilla. In contrast, interdental plates are large and quite distinct in other abelisaurids such as *Majungasaurus* (Sampson et al., 1996). As in other abelisaurids, the palatal process of the premaxilla is highly atrophied at this margin, without any medial projection.

Rahiolisaurus possesses the plesiomorphic theropod dental count of four premaxillary teeth (numbered here pm 1 to pm 4, counting from front to back), as in other abelisaurids (Smith, 2007). The teeth are conical, slightly compressed labiolingually, biconvex and teardrop-shaped in cross-section, with a sharp mesial carina but a rounded distal edge, where the long axis is oriented in the mesiodistal direction (Fig. 3.1f, g). Morphology of the premaxillary tooth varies from rostral to caudal position. Following the curvature of the rounded snout, the mesial carina in premaxillary tooth (pm) 1 is more lingually positioned but it gradually fades away in the caudal series (pm 4). The mesial carina is finely serrated in pm 1 where the density of serrations is 17 per 5 mm. In *Majungasaurus* (Smith, 2007), the serrations are somewhat coarser and more widely spaced (the average mesial serration density is about 12 per 5 mm). The estimated crown heights in premaxillary teeth are: pm 1 = 2.4 cm; pm 2 = 2.6 cm; pm 3 = 2.8; pm 4 = 2.4 cm. Passing backward, there is a gradual increase of the mesiodistal width. In *Majungasaurus* (Smith, 2007), the premaxillary teeth are more compressed labiolingually with stronger development of both the mesial and distal keels coarse serrations. However, these distal serrations are absent in the preserved teeth of *Rahiolisaurus*.

3.4.2 Vertebral Column

Presacral vertebrae count in articulated specimens of abelisaurids, such as *Carnotaurus* and *Majungasaurus* (Bonaparte et al., 1990; O'Connor, 2007), is 10 cervicals and 13 dorsals. Number of sacrals varies within Abelisauridae, being 5 in *Majungasaurus* (O'Connor, 2007) but 7 in *Carnotaurus* (Bonaparte et al., 1990). No less than 26 caudal vertebrae were present (O'Connor, 2007). In *Rahiolisaurus*, the vertebrae were found disarticulated and often jumbled with other bones.

3.4.2.1 Cervical Vertebrae

The axis (ISIR 658; Fig. 3.2a–c) was found in close association with cervical 3 (ISIR 659; Fig. 3.2d–f). The axis bears a distinctive, cone-like odontoid process on the rostral face of the centrum. Below the odontoid process there is a wide and deeply concave intercentral surface for reception of the atlantal centrum. The neural arch of the axis is transversely expanded and triangular in dorsal

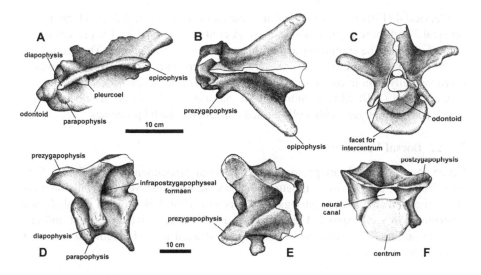

Fig. 3.2 Cervical vertebrae of *Rahiolisaurus gujaratensis*. (**a–c**) Axis (ISIR 658). (**d–f**) Cervical 3 (ISIR 659). (**a**) and (**d**) Left lateral, (**b**) and (**e**) dorsal, and (**c**) and (**f**) cranial views

view, because of wing-like, laterally-projecting postzygapophyses that terminate posteriorly in prominent epipophyses. The diapophyses are rod-like structures projecting ventrolaterally, similar to the condition in *Carnotaurus*, *Majungasaurus* and "*Compsosuchus*" (Huene and Matley, 1933; Bonaparte et al., 1990; Novas et al., 2004; O'Connor, 2007). A pneumatic opening is present, caudal to the diapophysis, and at least one pneumatic foramen is visible on the lateral surface of the centrum, as in these abelisaurids. The neural spine, though incomplete, is extensive with its dorsal edge caudally widened and V-shaped as seen in *Carnotaurus* and *Majungasaurus*.

Cervical 3 (ISIR 659; Fig. 3.2d–f) is the best preserved postaxial cervical vertebra in the collection; however, most of the postzygapophyses and neural spine are missing. The centrum is weakly amphicoelous and apparently devoid of pleurocoels, unlike other abelisauroids (e.g. *Carnotaurus*, *Majungasaurus*, *Laevisuchus*) in which double pleurocoels are present. The neural arch is transversely expanded and bears a prominent and straight ridge connecting both the pre- and postzygapophysis. The diapophyses are triangular in lateral view, but with a stout, rod-like ventral projection. A large pneumatic excavation lies at the base of the neural arch, immediately behind the diapophysis. The neural spine, broken at its base, is rectangular in cross-section, with its major axis transversely oriented. This transverse expansion of the neural spine resembles that of *Carnotaurus*, but it contrasts with the more laminar, craniocaudally extended condition present in cervical vertebrae of *Majungasaurus*. In ISIR 659, the prespinal depression is well excavated and wide, much more than in the mid-cervicals of *Majungasaurus* (O'Connor, 2007).

Cervical 4 (ISIR 660) was found in articulation with cervical 3. It is longer than cervical 3 and more angled in side view. A faint longitudinal keel is present on its ventral surface. The cranial surface of the centrum is subrectangular. The caudal rim of the centrum is more oval than the cranial surface. The diapophyses are directed more ventrally than in cervical 3, and lie closer to the parapophyses.

Cervical 6 (ISIR 515) is more robust than the previous one. It shows highly developed epipophyses with well-defined cranial and caudal processes.

3.4.2.2 Dorsal Vertebrae

Seven posterior dorsals are present in the collection (specimens ISIR 415, 504, 505, 506, 508, 509, 510). However, neural arches were not preserved, so available morphological information is restricted to the centra. ISIR 508 shows a spool-shaped centrum, with well-expanded cranial and caudal surfaces, which are flat and circular in outline. The centrum is strongly constricted at mid-length. No signs of pleurocoels are evident.

3.4.2.3 Sacral Vertebrae

Several associations of sacral vertebrae are preserved: specimens ISIR 506–507 and ISIR 516–517 consist of two fused elements and ISIR 511–514 and ISIR 404–407b consist of five fused elements. Most of the neural arches of these vertebrae are broken, and the sacral ribs are covered with hard matrix that obscures anatomical detail.

The most complete sacral series (ISIR 511–514 and 404–407b) show a series of concave profiles of the ventral margin of centra in side view (Fig. 3.3a, b). The centra are transversely constricted and lack pneumatic foramina. Comparison with sacral count of *Majungasaurus* (with a sacrum composed by five sacrals; O'Connor, 2007) and *Carnotaurus* (with a sacrum composed by seven sacrals; Bonaparte et al., 1990) suggests that ISIR 511–514 and 404–407b are represented by sacrals 1–5. Fused vertebrae of ISIR 516–517, on the other hand, may represent sacrals 5 and 6

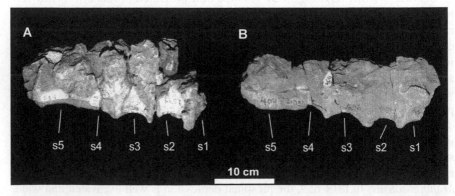

Fig. 3.3 Sacral vertebrae of *Rahiolisaurus gujaratensis*. (**a**) Right lateral view of sacrals (ISIR 511–514); (**b**) right lateral view of sacrals (ISIR 404–407b)

respectively, on the basis of their elongate and slender proportions. The remaining two fused sacral centra from the Rahioli collection (ISIR 506–507) are tentatively considered as sacrals 1 and 2 of a smaller individual.

Our interpretation of ISIR 511–514 and 404–407b is as follows: sacrals 1 and 2 are craniocaudally short and constricted at mid-length, thus their respective cranial and caudal rims are prominent; sacral 3 is more elongate than the preceding vertebra, in which its cranial end is wider than the caudal one (consequently, the ventral margin of sacral 3 is inclined upwards); sacrals 4 and 5 are spool-shaped and smaller in diameter than the preceding vertebrae, and their respective ventral margins form a wide arch in lateral aspect in continuation with the inclined ventral margin of sacral 3. The modest caudal diameter and elongate proportions of sacrals 4 and 5 suggest that ISIR 511–514 and 404–407b probably contained seven sacrals, as in *Carnotaurus*, thus contrasting with the less derived condition of five sacrals present in *Majungasaurus*. The relative space on the medial surface of ilia from the Rahioli collection indicates the incorporation of nine vertebrae for sacral attachment (last dorsal + 7 sacrals + caudal 1), a count that is also documented in *Carnotaurus*. Apparently, *Rahiolisaurus* shows a more marked ventral curvature and constriction of mid-sacrals than that of the noasaurid *Masiakasaurus* (Carrano et al., 2002), and the abelisaurids *Majungasaurus* (O'Connor, 2007), *Rajasaurus* (Wilson et al., 2003), and *Lametasaurus indicus* (Matley, 1923). However, *Rahiolisaurus* did not attain a high degree of ventral concavity and constriction of sacrals 2–7 as seen in *Carnotaurus* (Bonaparte et al., 1990). Moreover, sacrals of *Rahiolisaurus* are not transversely as narrow as in *Carnotaurus*.

3.4.2.4 Caudal Vertebrae

Forty-six caudal vertebrae are available in the collection (specimens ISIR 408–429 and ISIR 518–545). One isolated chevron was also found (ISIR 546). Except for ISIR 410, which represents a proximal caudal, most of the available vertebrae correspond to the mid- and distal portions of the tail. These vertebrae represent different individuals of different sizes and growth stages. Because of this ontogenetic variation, caudals with almost the same length often exhibit different morphologies, as expressed by the depth of the centra and development of the neural arches.

The proximal caudal (ISIR 408) exhibits a short centrum, which is slightly deeper than longer; its cranial articular surface is circular and inflated. The centrum is in the shape of a parallelogram in side view, with both cranial and caudal margins slightly inclined cranially, and the ventral margin concave. The caudal rim is more ventrally offset than the cranial one; this asymmetric condition suggests that this vertebra occupied a position close to the sacrum. The base of the neural arch is transversely wide and dorsoventrally low, a condition that contrasts with proximal caudals of *Carnotaurus* and *Pycnonemosaurus*, in which the base of the neural arches is transversely narrow and dorsoventrally deep. Transverse processes project laterally and slightly dorsally, in contrast to the condition in *Carnotaurus* and *Pycnonemosaurus* in which they are projected more dorsally. Transverse processes of ISIR 408 are robust at their bases, and triangular in cross-section. Because their distal ends are broken, it is not possible to infer if they were fan-shaped, a feature that occurs in most (e.g. *Carnotaurus, Ilokelesia, Ekrixinatosaurus*) but not all (e.g. *Majungasaurus*) abelisaurids. Yet, it is possible to recognize that both dorsal and cranial surfaces of transverse processes are deeply excavated, thus resulting in the development of a prominent cranial margin. This condition is also present in proximal caudals of Patagonian abelisaurids (e.g. *Aucasaurus; Ekrixinatosaurus novasi* MUCPv-294). However, ISIR 408 shows that such a cranial rim curves dorsally. The base of the transverse process is devoid of the buttresses and excavations present in proximal caudals of *Majungasaurus* (O'Connor, 2007) and *Aucasaurus* (Coria et al., 2002).

Mid-caudals exhibit neural arches that are triangular in dorsal aspect, with well-developed transverse processes. The centrum bears a reduced longitudinal ridge on its lateral surface, and its ventral surface is keeled. Both cranial and caudal articular surfaces of the centrum are deeply concave.

More distal caudals are characterized by a centrum that is polygonal in cross-section, due to the presence of strong longitudinal ridges that are separated from the remainder of the lateral surface of the centrum by a deep groove. The ventral surface of the centrum is flat, and the cranial and caudal articular facets are moderately concave. The neural arch is Y-shaped in dorsal view, conferred by the cranially projected, paired prezygapophyses, that are continued caudally by a narrow and elongate central portion of the neural arch.

The chevron bone (ISIR 546) is dorsoventrally elongate, with a caudally curved shaft.

3.4.3 Pectoral Girdle

A partial left scapulocoracoid (ISIR 465; Fig. 3.4a, b, c), the proximal half of a left scapula (ISIR 645), the proximal portion of a right scapula (ISIR 547), and two-thirds of a left scapula (ISIR 432; Fig. 3.4d, e) are preserved.

A left scapula (ISIR 432) shows the delicate cranial and caudal margin of the blade (Fig. 3.4d, e), which is narrow (8 cm in craniocaudal diameter, taken at the widest level of the blade) and elongate (approximately 45 cm proximodistally long, as preserved). In lateral view, the blade shows a gentle curvature, where the caudal margin is convex and the cranial margin is concave, a condition present in ISIR 645 and 465. In *Carnotaurus*, similar curvature of the scapula is present but its caudal margin exhibits a prominence that is separated from the glenoid cavity by a concave notch. Notably, such a prominence is associated with a ridge running longitudinally on the medial surface of the bone (Fig. 3.4d).

In *Rahiolisaurus*, the acromion process is incomplete; however, a subhorizontal rugosity, resembling that of *Carnotaurus* exists on its lateral surface. On the internal surface of scapula (ISIR 432), and lying close to the glenoid cavity, a sliver of bone, separated from the scapula by a layer of sediment, may correspond with an unknown ossification also reported in *Carnotaurus* (see Bonaparte et al., 1990, Fig. 27, labelled with a question mark).

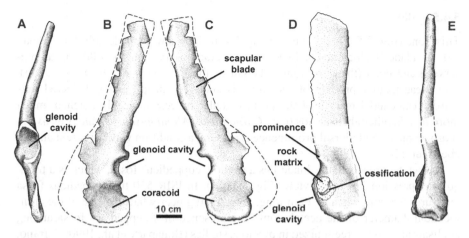

Fig. 3.4 Scapular girdle of *Rahiolisaurus gujaratensis*. (**a–c**) Left scapulocoracoid (ISIR 465); (**a**) caudal, (**b**) lateral, and (**c**) medial views. (**d** and **e**) Left scapular blade (ISIR 432); (**d**) medial, and (**e**) cranial views

On the basis of the development of the scapular and coracoidal lips of the glenoid (Fig. 3.4a, b, c), it seems that this cavity is oriented more caudally than in *Carnotaurus* and *Majungasaurus* (where the glenoid faces caudodorsally, arbitrarily assuming a vertical position for the entire scapulocoracoid). In relation with this feature, the glenoid cavity of ISIR 465 seems to have been separated from the caudodorsal margin of coracoid through a notch, a condition that is absent in *Carnotaurus*, *Aucasaurus* and *Majungasaurus*. The coracoid (represented in ISIR 465 by its caudal half), forms a broad and shallow oval plate.

3.4.4 Forelimbs

Chatterjee and Rudra (1996) referred some forelimb elements (e.g., right humerus and radius) from the Rahioli site to an abelisaur. However, the humerus in question (ISIR 434) is short, robust, and expanded at both ends; its proximal end lacks the typical abelisaurid features such as bulbuous head and highly reduced deltopectoral crest (Novas et al., 2006) and is interpreted here as belonging to a juvenile titanosaur. On the other hand, the purported right radius (ISIR 549) is in fact an abelisaurid metatarsal broken in half longitudinally. In conclusion, the forelimb elements were misidentified previously and are not present in the ISI collection.

3.4.5 Pelvic Girdle

The pelvic girdle is represented by five ilia fused with proximal ends of pubes (ISIR 436, 437, 634, 550, and 551), four pubes (ISIR 464, 552, 553 and 554), and two ischia (ISIR 439).

3.4.5.1 Ilium

This bone (Fig. 3.5a, b, c, d) is rectangular in side view, considerably longer than tall. Its blade is elongate (in ISIR 436, the craniocaudal length is 80 cm, as preserved) and deep (the dorsoventral height of the postacetabular blade is 22 cm). The preacetabular process of the ilium is short but deep, and is deflected outwards. The caudal margin of the postacetabular process is deeply notched, much more than in other abelisaurids (e.g., *Carnotaurus, Majungasaurus, Aucasaurus*). A sharp, conical caudodorsal prominence is present, a condition widely present among abelisauroids.

Below the preacetabular blade lies a shallow cuppedicus fossa, which is a triangular depression in ventral view (Fig. 3.5b, d). In ISIR 550 the cuppedicus fossa measures 5.5 cm wide and 7 cm long, representing nearly half the size of the craniocaudal diameter of the acetabulum. The presence of a cuppedicus fossa among abelisaurids was not recognized in previous studies (Bonaparte et al., 1990; Carrano,

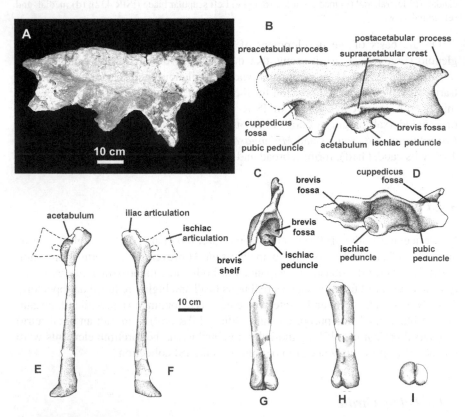

Fig. 3.5 Pelvic girdle of *Rahiolisaurus gujaratensis*. (**a–d**) Ilium of *Rahiolisaurus gujaratensis*, (**a–c**) left ilium (ISIR 436); (**a**) medial, (**b**) lateral, and (**c**) caudal views. (**d**) Right ilium (ISIR 550) in ventral view. (**e and f**) Right pubis (ISIR 554); (**e**) lateral and (**f**) medial views. (**g–i**) Fused ischia (ISIR 439); (**g**) cranial, (**h**) caudal, and (**i**) distal views

2007), though the feature appears to be present in *Majungasaurus* and *Lametasaurus indicus* (Matley, 1923).

The acetabulum is large, fully perforated, with a parabolic dorsal rim. At the dorsolateral roof of the acetabulum, the supracetabular crest forms a pronounced overhanging rim, in which the cranial half is slightly concave while the caudal portion is prominent, convex, and contiguous with the brevis shelf. The brevis fossa, widens and deepens caudally.

The cranial margin of the acetabulum terminates ventrally in a pronounced pubic peduncle. It is very deep, thus its dorsal corner occupies a higher position with respect to the supracetabular crest. As a result, the preacetabular blade in side view hides the pubic peduncle considerably. The articular surface of the pubic peduncle is cranioventrally oriented a feature also present in *Majungasaurus* (Sampson et al., 1998; Carrano, 2007). In all preserved specimens of ilia, pubic peduncles are strongly fused with the proximal end of pubes, but the suture between these two bones remains visible.

The ischiac peduncle is large and triangular in cross section. From the ischiac peduncle, a sheet of bone extends caudally to accommodate the brevis shelf.

3.4.5.2 Pubis

The pubis is a narrow, elongate, and distally footed (Fig. 3.5e, f). As in *Carnotaurus* and *Majungasaurus*, the pubis is fused to the pubic peduncle of the ilium. The proximal end of the pubis, which articulates with the ilium, is massive. Below the acetabulum, and cranial to the ischial contact, the pubis becomes a thin vertical plate that is perforated by a small obturator foramen, as it is seen in ISIR 552.

The proximal end of the pubis is acute in lateral aspect, directing at an angle of 45° between its cranial margin and its articular surface for the ilium. This morphology of the pubis correlates with the cranioventral orientation of the articular facet of the pubic peduncle of ilium. Although the morphology of the pubic peduncle of the ilium is known in other abelisaurids (*Majungasaurus*), it is not well known in *Carnotaurus sastrei* (MACN-CH 894), where this region of the ilium is partially broken. Thus, the pubic peduncle of *Rahiolisaurus* provides new information for this portion of the pelvis. Recently, Carrano (2007) described the unusual presence of a distinct peg in the ilium of *Majungasaurus* that presumably fitted into a corresponding socket on the pubis (a bone that is missing in *Majungasaurus*). In the noasaurid *Masiakasaurus* the proximal end of pubis also exhibits a socket for the reception of the peg-like projection of the pubic pedicle of ilium (Carrano et al., 2002). Wilson et al. (2003) also recognized a similar peg-like projection in the ilium of *Rajasaurus* for an opposing socket on the pubis. Such a peg-like process is represented, at least in ISIR 550 and 436, by a conical, ventral projection. In *Rahiolisaurus* this "peg" is separated demarcated from the pubic peduncle of the ilium through an outer rim.

The shaft of the pubis is transversely narrow and almost straight in side view. The most complete pubis in the collection (ISIR 554), lacking part of its distal foot and proximal end for iliac contact, measures 70 cm long; the total length of this pubis is estimated to be 75–80 cm when complete. The transverse diameter of the same

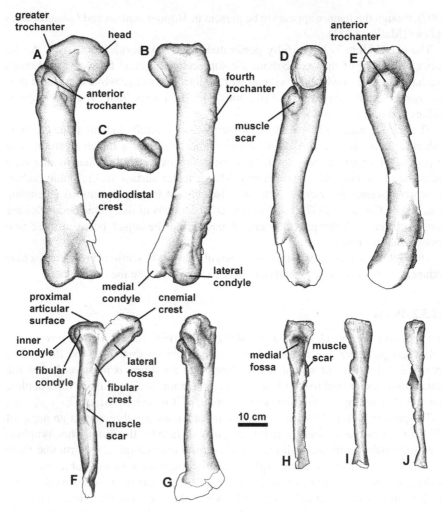

Fig. 3.6 Hindlimbs of *Rahiolisaurus gujaratensis*. (**a–e**) Right femur (ISIR 557); (**a**) cranial, (**b**) caudal, (**c**) proximal, (**d**) medial, and (**e**) lateral views. (**f and g**), right tibia (ISIR 445); (**f**) lateral, and (**g**) cranial views. (**h–j**) Left fibula (ISIR 555); (**h**) medial, (**i**) lateral, and (**j**) cranial views

bone at its distal quarter is 5 cm. The pubic symphysis extends most of the length of the bone, being interrupted above the distal pubic foot by an elliptical opening, as seen in ISIR 553.

3.4.5.3 Ischium

Available specimens of ischium (Fig. 3.5g–i) do not preserve their proximal halves for contact with ilium and pubis. The ischium is more robust than the pubis, as it also occurs in *Carnotaurus*. The cranial surface of the shaft is longitudinally concave;

its caudal surface is both longitudinally and transversely convex. The shaft flares slightly at the distal end where the distal expansion is more developed caudally than cranially. The two ischia are joined along a long median symphysis for a great length and are partially fused at their distal region. On the caudal surface of the ischia, a pyramidal prominence marks the point where both ischia fuse to each other up to their distal end.

3.4.6 Hind Limb

Hind limb elements are represented by eleven femora (ISIR 440, 441, 442, 443, 444, 556, 557, 558, 559, 560, and 561), ten tibiae (ISIR 438, 445, 446, 449, 457, 458, 562, 563, 564, and 565), three tibiae fused with their respective astragalus and calcaneum (ISIR 447, 570 and 571), a fibula (ISIR 555), metatarsals (ISIR 573, 450, 475, 569, 573, 575, 581 and 586), and pedal phalanges (ISIR 576–579, 582–585, and 587–595). In some associations, fusions between the tibia and astragalus and between metatarsals II and III are observed.

3.4.6.1 Femur

Available femora belong to at least five different individuals corresponding to different size classes. Aside from differences in size and proportion, all the femora in the collection show similar morphology. They share many features with those of *Carnotaurus*, *Ekrixinatosaurus*, and *Xenotarsosaurus* including a rounded and proportionally large femoral head, a low but blunt anterior trochanter contiguous with a trochanteric shelf, a reduced fourth trochanter, and a pronounced and longitudinally extended crest along the craniomedial edge of the distal shaft (Fig. 3.6a–e). ISIR 557 represents the best preserved femur in the collection. The bone is transversely wide in relation to its length, where the width is about 12–16% of the length of the bone. The femoral head is very large, about 20% of the whole length of the bone. The anterior trochanter projects cranially as a distinct flange. However, it lies far below the greater trochanter. A prominent trochanteric shelf is present. Farther downward, the shaft shows a greatly enlarged mediodistal crest, a synapomorphy for abelisauroids (Sampson et al., 2001).

3.4.6.2 Tibia

Available tibiae are shorter than the longest femora: the difference in length between the longest tibia and the longest femora is 15 cm. In other words, the longest tibia is about 30% shorter than the longest femur. In some cases the tibia is not fused with the astragalus, in others, these two bones are fused to form a tibiotarsus. Apparently, the fusion of the tibiotarsus is not correlated with the absolute size of the specimens. For example, a 60-cm-long tibia (ISIR 445) lacks astragalar fusion, whereas a 43-cm-long tibia (ISIR 438), belonging to a relatively smaller individual, shows such fusion. Whether this reflects true dimorphism or whether it implies a result of individual or taxonomic variation is not clear. However, the tibiotarsus (ISIR 571) looks more robust when compared with a similar-sized tibia (ISIR 445).

The tibia is a stout bone (Fig. 3.6f, g). The cnemial crest is large and projects considerably both cranially and dorsally, a feature characteristic of abelisauroids (Sampson et al., 2001). At mid-length of the lateral surface of the tibial shaft there is an elliptical rugosity, a prominent muscle scar bordered by a rim. This feature is also present in the Patagonian abelisaurid *Ekrixinatosaurus novasi* (MUCPv-294). The tibial shaft is craniocaudally compressed and transversely wide. The cranial surface of the distal end is concave to receive the ascending process of the astragalus. However, there are no ridges to overhang on the ascending process of the astragalus.

3.4.6.3 Fibula

It is a slender and elongate bone with proximal and distal ends expanded in a craniocaudal direction (Fig. 3.6 h, i, j). The cranial surface of the proximal end is transversely flat and wide, being separated from the medial surface of the bone through a sharp longitudinal ridge, which is medially offset at the level of the medial fossa, which probably was the site of origin of part of the pedal flexor musculature. The medial fossa is triangular, opens caudally and extends down the fibular shaft. At the lower level of the fossa, a prominent, quadrangular process emerges from the cranial margin of the bone for insertion of *M. iliofibularis*. The medial surface of the fibular shaft is longitudinally grooved.

3.4.6.4 Tarsus

The astragalus and calcaneum are firmly fused together. The calcaneum would be proportionally wide, with a maximum width of roughly 5 cm, thus approximately representing 36% of the maximum transverse width of the astragalus (14 cm). The ascending process is rectangular, roughly 5.5 cm deep, representing almost the same depth as the astragalar body. Its cranial surface exhibits a single horizontal groove.

3.4.6.5 Metatarsals and Pedal Phalanges

ISIR 569 is the best-preserved metatarsal II in the collection (Fig. 3.7a, b, c). The proximal end is narrow transversely but the distal end is inflated. Except for the distal ginglymoid joint surface, all of the lateral surface of this bone is flat and served for articulation with metatarsal III. Immediately proximal to the distal end, the cranial margin of the shaft develops a slight flange, which projects cranially. Such a flange is also present in metatarsal II of *Masiakasaurus knopfleri* (Carrano et al., 2002), *Noasaurus leali* (PVL 4370), and *Rajasaurus narmadensis* (Wilson et al., 2003). Notably, metatarsal II of *Rahiolisaurus* is considerably narrower proximally than that of *Rajasaurus narmadensis* (Wilson et al., 2003).

Metatarsal III (Fig. 3.7d, e) is almost straight in cranial aspect but slightly curved in side view (caudal convexity). Its proximal end is much more voluminous than the

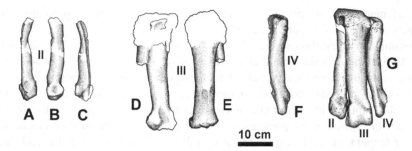

Fig. 3.7 Metatarsals of *Rahiolisaurus gujaratensis*. (**a–c**) Right metatarsal II (ISIR 569); (**a**) dorsal, (**b**) lateral, and (**c**) plantar views. (**d** and **e**) Left metatarsals II and III (ISIR 573); (**d**) dorsal and (**e**) plantar views. (**f**) Left metatarsal IV (ISIR 475) in dorsal view. (**g**) Composite reconstruction of metatarsals II–IV

proximally slender metatarsals II and IV. The caudal surface of the proximal end is blunt, forming a structure superficially resembling the avian hypotarsus, a feature noted for the abelisaurid *Aucasaurus* (Coria et al., 2002). Distal to this projection, the caudal surface exhibits two strong longitudinal ridges that converge distally before reaching the distal end. These two ridges separate the caudal surface from the lateral and medial ones to form the ginglymoid joint. The internal surface of the proximal end of metatarsal III is deeply excavated for the reception of metatarsal II. A craniomedial longitudinal ridge is developed distally, where it forms a weak, cranially-projecting flange. When articulated, metatarsal II lies closely appressed against metatarsal III (Fig. 3.7g), and the distal flange developed on the craniomedial margin of metatarsal III matches, in shape and position, with a similar prominence on metatarsal II.

Metatarsal IV is more slender and shorter than metatarsal III and the shaft is bowed somewhat laterally (Fig. 3.7f, g).

There is some indication that a splint-like element, interpreted here as metatarsal I, may be present in ISIR 573 in association with metatarsals II and III. Further preparation of the ISIR 573 is required to reveal the full anatomy of this bone. The rod-like bone extends from the proximal end of metatarsal III up to its mid-length. Metatarsal I of *Rahiolisaurus* seems to be more delicate and reduced than in *Aucasaurus* (Coria et al., 2002).

3.5 Discussion and Conclusion

Although Chatterjee and Rudra (1996) referred the new abelisaurid material from the Rahioli site to *Indosuchus raptorius*, the current study suggests that the assorted material described here belongs to a new taxon, *Rahiolisaurus gujaratensis*, represented by several individuals of different growth stages. It is estimated that an adult individual of *Rahiolisaurus* (ISIR 557, femur length 77 cm) was a large-sized abelisaurid (probably 8 m long), as calculated from *Majungasaurus crenatissimus*, with a femur length of >56.8 cm, and an estimated whole length of approximately 6 m (Krause et al., 2007; Carrano, 2007)

Rahiolisaurus exhibits several characteristics that are present in Abelisauroidea and Abelisauridae: axial neural arch transversely expanded and triangular in dorsal view; postaxial cervical vertebrae with prominent ridge connecting both pre- and postzygapophysis; broad development of pre- and postspinal fossae in postaxial cervicals; scapula with caudal margin convex and cranial margin concave (a feature shared with *Carnotaurus*); presence of distinct, elliptical ossification on medial surface of scapula, at level of glenoid cavity (also in *Carnotaurus*); pubis and ilium tightly fused; postacetabular blade of ilium bearing sharp caudodorsal prominence; pubis transversely narrow and distally footed; ischia massive and partially fused to other along symphysis.

It is difficult to evaluate the taxonomic status and validity of Indian abelisaurids, as most species have been established on the basis of fragmentary remains. This is the case for the abelisaurids *Lametasaurus indicus* (Matley, 1923), *Indosuchus raptorius* (Huene and Matley, 1933; Chatterjee, 1978), and *Indosaurus matleyi* (Huene and Matley, 1933; Chatterjee, 1978). These Indian abelisaurids have had a complex and checkered taxonomic history, already reviewed in several recent papers (see Wilson et al., 2003; Novas et al., 2004; Carrano and Sampson, 2008). Unfortunately, the holotypic materials supporting each of these taxa do not offer clear autapomorphic features (Novas et al., 2004; Carrano and Sampson, 2008), and it is difficult (if not impossible) to refer the several isolated bones collected in the same quarry to any of the above mentioned species. Moreover, many of the specimens originally described by Matley (1923) and Huene and Matley (1933) are currently lost. The specimens of *Rahiolisaurus*, although abundant and well preserved, do not offer enough information that help to resolve the taxonomic validity of the abelisaurid mentioned above.

Because of these problems, we believe that *Lametasaurus indicus*, *Indosuchus raptorius*, and *Indosaurus matleyi* must be considered as nomina dubia.

More recently, Wilson et al. (2003) erected *Rajasaurus narmadensis* on the basis of associated remains of a braincase and some postcranial bones from the Temple Hill locality, about 400 m northwest of our Rahioli site. This taxon is based on associated cranial and postcarnial elements, bearing autapomorphic features in the temporal region of the skull (Wilson et al., 2003). So, we opt to recognize, at the moment, two abelisaurid species from the Lameta Formation: *Rajasaurus narmadensis* and *Rahiolisaurus gujaratensis*.

Comparisons between *Rahiolisaurus* with *Rajasaurus* are restricted to some common elements (i.e., sacrum, ilium, distal femur, fibula, metatarsal II). The sacral series of *Rahiolisaurus* shows a more pronounced ventral curvature and constriction of mid-sacrals than the sacrum of *Rajasaurus* (Wilson et al., 2003). The hind limbs of *Rahiolisaurus* are more slender than those of *Rajasaurus*; the ilium of *Rahiolisaurus* is quite distinctive compared to that of *Rajasaurus* with the development of a deep ventral notch on the postacetabular process; its dorsal acetabular rim is parabolic, whereas in *Rajasaurus*, this rim is circular. Moreover, metatarsal II is highly distinctive in being much narrower proximally in *Rahiolisaurus* than in *Rajasaurus*. *Rajasaurus* appears to represent heavy-bodied, stout-limbed abelisaurids (Wilson et al., 2003), whereas *Rahiolisaurus* is a more gracile and slender-limbed form.

Comparisons of *Rahiolisaurus gujaratensis* with *Majungasaurus crenatissimus* from Madagascar (Sampson et al., 1998; Sampson and Krause, 2007) show some important distinctions. For example, cervical neural spines of *Rahiolisaurus* are transversely expanded, as in *Carnotaurus*, but different from the more plesiomorphic, blade-like ones of *Majungasaurus*. Moreover, *Rahiolisaurus* seems to be more closely related with forms that are more derived than *Majungasaurus*, as evidenced by the relatively large number of sacral vertebrae. *Rahiolisaurus* had seven co-ossified sacrals as in *Carnotaurus*, whereas *Majungasaurus* possessed only five. Although Madagascar and India were palaeogeographically close at the end of the Cretaceous, *Rahiolisaurus* does not exhibit close resemblance to the contemporary *Majungasaurus*. Although *Rahiolisaurus* resembles *Carnotaurus* in many postcranial features and shares a close ancestry, the lack of skull roof precludes conclusions on whether or not *Rahiolisaurus* was a derived member of the horned abelisaurid clade Carnotaurinae.

It appears from the discussion that *Rahiolisaurus* is distinct from *Rajasaurus, Carnotaurus*, and *Majungasaurus* and shows several autapomorphies in the structure of ilium and metatarsals that warrant erection of a new taxon.

Phylogenetic relationships of *Rahiolisaurus* within Abelisauridae need to be evaluated in the context of a cladistic analysis of the entire family, a task that is outside the scope of the present paper.

Acknowledgments The palaeontological fieldwork in India was carried out in collaboration with Texas Tech University, the Indian Statistical Institute, and the Geological Survey of India. We thank the National Geographic Society, the Dinosaur Society, and the Smithsonian Institution for continued funding of the field projects in India and the Directors of the Indian Statistical Institute and the Geological Survey of India for logistical support. We thank D. Pradhan and Shyamal Roy for field assistance, the Indian Statistical Institute and the Geological Survey of India for access to the theropod collections. We thank Catherine Forster, Dave Krause, José Bonaparte, Alejandro Kramarz, Jorge Calvo and Rodolfo Coria for access to specimens under their care, Jorge González and Jeff Martz for illustrations, and Bill Mueller for photography. Thanks are given to Bill Mueller, Jeff Wilson, Sara Burch, Dave Krause, Matt Carrano and Saswati Bandyopadhyay for their useful comments on the manuscript. Texas Tech University, Agencia Nacional de Promoción Científica y Técnica, CONICET, Fundación Antorchas, National Geographic Society, and The Jurassic Foundation supported the research.

References

Bonaparte, JF, Novas, FE, Coria, RA (1990) *Carnotaurus sastrei* Bonaparte, the horned, lightly built carnosaur from the Middle Cretaceous of Patagonia. Nat Hist Mus L A County, Contr Sci, 416:1–42.

Carrano, MT (2007) The appendicular skeleton of *Majungasaurus crenatissimus* (Theropoda: Abelisauridae) from the Late Cretaceous of Madagascar In: Sampson, SD, Krause, DW (eds) *Majungasaurus crenatissimus* from the Late Cretaceous of Madagascar, Soc Vertebr Paleontol Mem, 8:163–179.

Carrano, MT, Sampson, SD (2008) The phylogeny of the Ceratosauria (Dinosauria: Theropoda). J Syst Palaeontol, 6:183–236.

Carrano, MT, Sampson, SD, Forster, CA (2002) The osteology of *Masiakasaurus knopfleri*, a small abelisauroid (Dinosauria: Theropoda) from the Late Cretaceous of Madagascar. J VertPaleontol, 22:510–534.

Chatterjee, S (1978) *Indosuchus* and *Indosaurus*, Cretaceous carnosaurs from India. J Paleontol, 52:570–580.

Chatterjee, S, Rudra, D (1996) KT events in India: impact, rifting, volcanism and dinosaur extinction. In: Novas, FE, Molnar, RL (eds) Proceedings of the Gondwanan Dinosaur Symposium. Brisbane, QLD, Mem Queensland Mus, 39(3):489–532.

Coria, R, Chiappe, LM, Dingus, L (2002) A new close relative of *Carnotaurus sastrei* Bonaparte, 1985 (Theropoda: Abelisauridae) from the Late Cretaceous of Patagonia. J Vertebr Paleontol, 22:460–465.

Huene, FV, Matley, CA (1933) The Cretaceous Saurischia and Ornithischia of the central provinces of India. Mem Geol Surv Ind (Palaeontol Indica), 21:1–74.

Krause, DW, Sampson, SD, Carrano, MT, O'Connor, PM (2007) Overview of the history of discovery, taxonomy, phylogeny, and biogeography of *Majungasaurus crenatissimus* (Theropoda: Abelisauridae) from the Late Cretaceous of Madagascar. In: Sampson, SD, Krause, DW (eds) *Majungasaurus crenatissimus* (Theropoda: Abelisauridae) from the Late Cretaceous of Madagascar, Soc Vertebr Paleontol Mem, 8:1–20.

Matley, CA (1923) Note on an armoured dinosaur from the Lameta beds of Jubbulpore. Rec Geol Sur Ind, 55:105–109.

Novas, FE, Agnolín, FL, Bandyopadhyay, S (2004) Cretaceous theropods from India: a review of specimens described by Huene and Matley (1933). Rev Mus Argentino Cienc Nat "Bernardino Rivadavia", 6:67–103.

Novas, FE, Bandyopadhyay, S (1999) New approaches on the Cretaceous theropods from India. Proceedings of the 7th International Symposium on Mesozoic Terrestrial Ecosystems. Aso Paleontol Argentina, Publ Esp, Buenos Aires, Argentina, 7:46–47.

Novas, FE, Bandyopadhyay, S (2001) Abelisaurid pedal unguals from the Late Cretaceous of India. Aso Paleontol Argentina Publ Esp, 7:145–149.

Novas, FE, Ezcurra, M, Agnolin, F (2006) Humerus of a basal abelisauroid theropod from the Late Cretaceous of Patagonia. Rev Mus Argentino Cienc Nat "Bernardino Rivadavia", 8:63–68.

O'Connor, P (2007) The postcranial axial skeleton of *Majungasaurus crenatissimus* (Theropoda: Abelisauridae) from the Late Cretaceous of Madagascar. Society of Vertebrate Paleontology Memoir 8. J Vert Paleontol, 27:127–162.

Sampson, SD, Carrano, MT, Forster, CA (2001) A bizarre predatory dinosaur from the Late Cretaceous of Madagascar. Nature, 409:504–506.

Sampson, SD, Krause, DW (2007) *Majungasaurus crenatissimus* (Theropoda: Abelisauridae) from the Late Cretaceous of Madagascar. Soc Vert Paleontol Mem, 8:1–184.

Sampson, SD, Krause, DW, Dodson, P, Forster, CA (1996) The premaxilla of *Majungasaurus* (Dinosauria: Theropoda), with implications for Gondwana paleobiogeography. J Vertebr Paleontol, 16:601–605.

Sampson, S, Witmer, L, Forster, C, Krause, D, O'Connor, P, Dodson, P, Rovoavy, F (1998) Predatory dinosaur remains from Madagascar: implications for the Cretaceous biogeography of Gondwana. Science, 280:1048–1051.

Sahni, A, Bajpai, S (1991) Eurasiatic animals in the Upper Cretaceous nonmarine biotas of peninsular India. Cret Res, 12:177–183.

Sereno, PC, Wilson, JA, Conrad, JL (2004) New dinosaurs link southern landmasses in the Mid-Cretaceous. Proc R Soc Lond, B271:1325–1330.

Smith, JB (2007) Dental morphology and variation in *Majungasaurus crenatissimus* (Theropoda: Abelisauridae) from the Late Cretaceous of Madagascar. In: Sampson, SD, Krause, DW (eds) *Majungasaurus crenatissimus* (Theropoda: Abelisauridae) from the Late Cretaceous of Madagascar, Soc Vert Paleontol Mem, 8:103–126.

Wilson, JA, Sereno, PC, Srivastava, S, Bhatt, DK, Khosla, A, Sahni, A (2003) A new abelisaurid (Dinosauria, Theropoda) from the Lameta Formation (Cretaceous, Maastrichtian) of India. Contr Mus Paleontol Univ Michigan, 31:1–42.

Chapter 4
Pterosauria from the Late Triassic
of Southern Brazil

J.F. Bonaparte, C.L. Schultz, and M.B. Soares

4.1 Introduction

Triassic pterosaurs are recorded mostly from Norian beds of northern Italy, represented by almost complete specimens, which afford a glimpse of the evolutionary level reached by them at that time. *Eudimorphodon* (Zambelli, 1973), *Peteinosaurus* (Wild, 1978) and *Preondactylus* (Wild, 1983) exhibit anatomical features typical of active flyers in the pectoral girdle and fore limbs, very similar to pterosaurs from the Jurassic. This situation allowed the interpretation by Wild (1978), Wellnhofer (1991), and others that the early differentiation of pterosaurs from their ancestors happened well before the Late Triassic, perhaps during the Late Permian (Wild, 1978), an assumption not supported by von Huene (1914, 1956) or Padian (1984).

The few pterosaur remains from the Late Triassic of Brazil described here show several more primitive characters than those from the European Norian, indicating that derived and primitive pterosaurs lived at almost the same time in different geographic settings (Jenkins et al., 2001), and that the early evolution of the group might have been faster than estimated.

The few and incomplete pterosaur remains discussed here represent the first evidence of the presence of the group in the Late Triassic of South America. Unfortunately the material is meagre and permits only a few tentative insights on the early evolutionary history of this interesting group of flying archosaurs.

A few fossil remains tentatively referred to Pterosauria were discovered during field work carried out during 2002 and 2005 in an old quarry near the city of Faxinal do Soturno, RGS, in southern Brazil. The material was found within a single small block of sandstone and considered to correspond to a single individual because of their size and structure. Their hollow features and the small proportional thickness of the wall of the diaphysis fit well with the interpretations by van der Meulen and

J.F. Bonaparte (✉)
Division Paleontología, Fundación de Historia Natural "Félix de Azara",
Buenos Aires 1405, Argentina
e-mail: bonajf@speedy.com.ar

S. Bandyopadhyay (ed.), *New Aspects of Mesozoic Biodiversity*,
Lecture Notes in Earth Sciences 132, DOI 10.1007/978-3-642-10311-7_4,
© Springer-Verlag Berlin Heidelberg 2010

Padian (1992) on such characters, recorded only in some birds and pterosaurs. Some of these bones, such as the proximal portion of the humerus and an almost complete coracoid, provided confident evidence of a primitive pterosaur.

The evolutionary history of this group of archosaurs is recorded since the Late Triasasic dolomites of northern Italy, probably of Middle to Late Norian age (Zambelli, 1973; Wild, 1984; Wellnhofer, 1991) and the Late Triassic from East Greenland (Jenkins et al., 2001). They are the oldest known reliably identified representatives of the Pterosauria, but show derived characters present in later pterosaurs, constituting three different families (Wild, 1978, 1984; Wellnhofer, 1991). Such evidence suggests that a significant earlier stage of their evolution is unknown.

The fossil remains discussed here have a special palaeontological interest because they are of similar or even older age than the most primitive pterosaurs from northern Italy. In addition, they were found in rocks that reflect a continental palaeoenvironment, different from the littoral marine environment of the Italian dolomites, and possibly more similar to the environment of the Greenland pterosaur (Jenkins et al., 2001). Such evidence suggests that at least part of the evolution of basal pterosaurs may have happened in continental or Mediterranean environments, as is the case for the Triassic sequence of southern Brazil, and that the colonization of the littoral marine realm was a further step.

Institutional abbreviations. UFRGS. PV – Department of Geology, Universidade Federal de Rio Grande do Sul, Porto Alegre, Brazil.

4.2 Systematic Palaeontology

Pterosauria Kaup 1834
Family indet
Faxinalipterus nov. gen

Diagnosis: Pterosaur with the following combination of characters: fibula not fused to the tibia, of the same length as the tibia, and with a distal expansion. Humerus with the major tuberosity higher than the humeral head; coracoid with a modest acrocoracoid process.

Etymology: *Faxinal*, after the city of Faxinal de Soturno, near which is the fossil locality, and *pterus*: wing.

Faxinalipterus minima nov. sp.

Diagnosis: Same as for the genus.

Holotype: UFRGS PV0927T includes a left coracoid lacking the sternal end; a proximal portion of the left humerus, a medial and distal fragment of the right humerus?; proximal fragments of left radius and ulna; an almost complete left femur; almost complete left tibia and fibula; fragments of right tibia and fibula associated with a possible metatarsal, and a few indeterminate fragments.

Etymology: *Minima*, after the small size of the holotype.

Referred material: UFRGS PV0769T, represented by most of a left maxilla with three teeth.

Geographic and stratigraphic provenance: Old quarry located 1.5 km NE of the city of Faxinal do Soturno, Rio Grande do Sul, Brazil. Beds of fine, massive sandstone from the lower section of the Caturrita Formation, early Coloradian (Rubert and Schultz, 2004; Bonaparte, 1973), approximately equivalent to the Carnian-Norian boundary of the European geochronology based on its content of derived cynodonts, dicynodonts and proterochampsids (Rubert and Schultz, 2004).

4.2.1 Description

The assignment of the holotype to Pterosauria is primarily based on the saddle-shaped morphology of the head of the humerus, on the general morphology of the coracoid, and the hollowed structure of the hollowed appendicular bones which bear very thin diaphyseal walls. The ratio between the thickness of the wall of the diaphysis and the diameter of the diaphyses is 1.5–7 up to 2.0–7. In the fragment of humerus that ratio is larger, 3.0–9, which agrees with van der Meulen and Padian (1992) indicating the humerus as the most robust appendicular bone in pterosaurs, especially in comparison with the hind limb bones.

4.2.1.1 Coracoid

Most of the possible left coracoid is preserved, lacking the sternal portion (Figs. 4.1a and 4.2a). Preserved length is 6.9 mm. The isolated condition of the coracoid suggests that it was not fused to the scapula. The proximal portion of the coracoid bearing the glenoid depression and the contact for the scapula forms an angle of about 130°.

Campylognathoides liasicus (Wellnhofer, 1974, fig. 6A) shows a similar angle. Near the area of contact with the scapula there is a small foramen tentatively identified as the coracoid foramen. On the angle formed by both portions of the coracoid there is a modest acrocoracoid process, less pronounced than in the Norian *Eudimorphodon* and *Peteinosaurus* (Wild, 1978, figs. 15 and 34).

4.2.1.2 Humerus

The proximal fragment of the left humerus represents approximately 1/3 of its estimated length (Figs. 4.1b and 4.2b). The length of the preserved portion is 6 mm. The saddle-shaped morphology of the humerus head is one of the most significant reasons to interpret *Faxinalipterus* as a pterosaur. The humeral head is transversely concave and ventrodorsally convex, bearing distally a distinct step separating it from the diaphysis. Medial to the head of the humerus is a distinct major tuberosity that is proximally higher than the head of the humerus. The deltoid crest is not seen in dorsal view. Possibly it is located ventrally and was not so developed as in pterosaurs from the Norian and Liassic of Europe (Zambelli, 1973; Wild, 1978; Wellnhofer, 1991). The transverse section of the diaphysis shows the hollowed character of the bone, with a thicker wall than in the hind limb bones described below, which is

Fig. 4.1 *Faxinalipterus minima.* The more significant pieces of the type specimen, UFRGS.PV0927T. **(a)** Distally incomplete left coracoid in postero-lateral view; **(b)** proximal half of the left humerus in dorsal view; **(c)** proximal half of left radius and ulna in medial view; **(d)** left femur in anterior view; **(e)** left tibia and fibula in posterior view; and **(f)** incomplete right tibia associated with a possible metatarsal. Measurements in the description

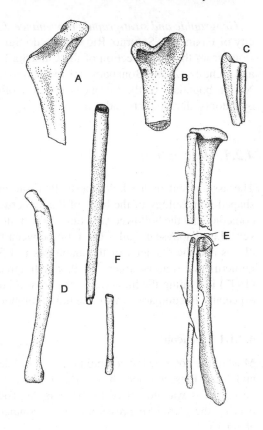

another diagnostic character of early pterosaurs. The medial and distal fragments of the right humerus? do not afford any significant information.

4.2.1.3 Ulna and Radius

Only the proximal portion of the left ulna and radius are preserved, possibly representing 1/3 of their total length (Figs. 4.1c and 4.2c). The preserved length is 7.5 mm. They are exposed in medial view; the ulna partially on the radius. The ulna has a concave glenoid for the humerus and a modest olecranon process similar to that of *Preondactylus* (Wild, 1983, fig. 3). The thickness of the ulna diminishes distally from the articular area. As preserved the diaphysis of the ulna is transversely compressed, but not the articular area that seems to preserve its original morphology. The articular area of the radius is proximally convex and transversely wide. In medial view most of the diaphysis of the radius is exposed and it is substantially thinner than the ulna. In some Triassic pterosaurs such as *Eudimorphodon* (Wild, 1978, Plates 2, 5, 8) and *Peteinosaurus* (Wild, 1978, Plate 14, fig. 36), the radius is proportionally thicker than in *Faxinalipterus*, a difference that increased in Liassic pterosaurs such as *Dorygnathus* (Wild, 1971; Padian and Wild, 1992), *Dimorphodon* (Padian, 1983), *Campylognathoides* (Wellnhofer, 1974), and even

Fig. 4.2 *Faxinalipterus minima.* UFRGS.PV0927T. (**a**) Left coracoid; (**b**) proximal half of the left humerus; (**c**) proximal half of left radius and ulna; (**d**) left femur; (**e**) left tibia and fibula; (**f**) incomplete right tibia with a possible metatarsal and (**g**) referred left maxilla. Measurements in the description

more in post-Liassic forms. It seems that *Faxinalipterus* has the more primitive condition of the thickening of the radius among pterosaurs.

4.2.1.4 Femur

The left femur is complete but with some portion broken (Figs. 4.1d and 4.2d). It is exposed in anterior and dorsal view, indicating some slight lateral bowing. Its total length is 23.3 mm, and the diameter of the diaphysis is 1.7 mm on its central area.

The head of the femur is partially damaged. It is small and clearly dorso-medially projected, more similar to *Dimorphodon* (Padian, 1983, fig. 7) than to *Eudimorphodon* (Wild, 1978, fig. 22). Near the femoral head and lateral to it is the trochanter externus (sensu Wild, 1978, p. 214), in a position and with morphology similar to that of *Eudimorphodon* (Wild, 1978) and *Dorygnathus* (Padian and Wild, 1978, fig. 22). In lateral view the distal area of the femur is thicker, but the condyles are not exposed.

4.2.1.5 Tibia and Fibula

The left tibia and fibula are almost complete, with some cracks that permit observations of the thickness of the diaphyseal wall (Figs. 4.1e, f and 4.2e, f). The right tibia and fibula are represented only by a small portion of the proximal area. Because both tibiae and fibulae are in the normal position relative to one another, possibly some strong ligaments or incipient ankylosis was developed in this primitive pterosaur.

The fibula does not have any longitudinal reduction, and its distal end bears a slight expansion. There is no indication of any ankylosis between these bones neither proximally nor distally. The length of the tibia is 26.4 mm and the fibula 25.2 mm. The thickness of the wall of the diaphysis somewhat above the medial length is approximately 0.12 mm, and that of the fibula somewhat less than 0.10 mm.

The proximal area of the fibula forms a distinct "articular head", somewhat projected backwards. The anterolateral trochanter is located on the proximal 1/5 of the fibular length.

The proximal area of the tibia has a rather large articulation for the femur, much larger than the proximal area of the diaphysis. Such an articular area overhangs the diaphysis on the lateral and posterior sides of the tibia. There are no indications of the areas on which the femoral condyles articulated.

The diaphysis of the tibia is subcircular along most of its length, except on the distal portion where it is less circular, with some indication of flattening. There is no indication of proximal tarsal bones incorporated with the tibia.

4.2.1.6 Metatarsal?

Among the collected material of this individual there are some fragmentary bones difficult to interpret properly (Figs. 4.1f and 4.2f). Some fragments lying on the sandstone in their original position, near an incomplete diaphysis of the right tibia? possibly correspond to incomplete metatarsals. One of them appears to be more or less complete, is 9.2 mm in length, with a diameter of approximately 0.5 mm. Its proximal area is broken, whereas the convex distal end may be complete.

4.2.1.7 Left Maxilla Referred to *Faxinalipterus*

An almost complete left maxilla bearing three teeth, UFRGS PV0769T, was collected from the same beds of sandstones and the same locality where the holotype of *Faxinalipterus* was found, but within a separate block of sandstones (Fig. 4.2 g). So it is not possible to interpret it with certainty as part of the holotype. The evidence that it may correspond to a pterosaur is rather weak. However, its small size, the morphology and the characters of the teeth do not preclude such a possibility.

The preserved portion of the maxilla may represent more than 90% of the complete bone. A fracture on its anterior margin suggests that the maxilla continued a little more anteriorly. The length of the preserved maxilla is 13 mm. It has a distinct dorsal process to contact with the nasal, indicating the posterior margin of the nasal opening. On the internal side of this area of the maxilla the medial process is preserved. The horizontal ramus has the dorsal and ventral margins parallel, showing little difference in height, which is approximately 1.3 mm.

All along the posterior margin of the dorsal process of the maxilla a step marks the anterior margin of the preorbital opening, a feature that is not present on the horizontal ramus of the maxilla.

Two of the three preserved teeth are complete. They are long and thin, a little bent posteriorly, 0.3 mm in diameter at the base and 1.0 mm in length. The approximate number of alveoli (this area is not fully prepared), is 18.

4.3 Comparisons and Discussion

Faxinalipterus is possibly the most primitive and oldest known pterosaur. Its stratigraphic level is considered late Ischigualastian or early Coloradian (Bonaparte, 1973; Rubert and Schultz, 2004), approximately equivalent to the Carnian-Norian boundary of the European geochronology. The presence of the procolophonian *Soturnia calodion* (Cisneros and Schultz, 2003), the dicynodont *Jachaleria candelariensis* (Araújo and Gonzaga, 1980), the basal trithelodontid *Riograndia guaibensis* (Bonaparte et al., 2001), and the basal saurischian *Guaibasaurus candelariensis* (Bonaparte et al., 1999), suggests a Triassic age older than the late Norian. Therefore, the Brazilian pterosaur may be a few million years older than the oldest European pterosaurs, which are from the middle or upper Norian beds of northern Italy (Wellnhofer, 1991).

Comparisons of the proximal portion of the humerus show the typical saddle-shaped morphology of the head of the humerus in pterosaurs. However, the normally pronounced deltopectoral crest present in all pterosaurs has not been observed in *Faxinalipterus*, perhaps because it is twisted underneath the preserved humerus, or perhaps it lacks the notable development present in more derived pterosaurs. The humerus of *Faxinalipterus* is the most robust piece among the available bones, a character that is diagnostic of the postcranium of primitive pterosaurs.

Usually the coracoid is fused to the scapula in most pterosaurs, as a functional requirement derived from muscular pressure necessary for flying, but not in *Faxinalipterus*. Possibly *Faxinalipterus* represents the primitive condition, and equally possibly it may be an ontogenetic character recorded in some pterosaurs as suggested by Wellnhofer (1991). The angle formed by the area of the coracoid bearing the glenoid and the ventrodistal portion to contact the sternum is typical of pterosaurs, except that the acrocoracoid process is normally more developed in the Late Triassic *Eudimorphodon* and *Peteinosaurus* (Wild, 1978, figs. 15 and 34) and in the Liassic pterosaurs. The poor definition of both the glenoid cavity and the acrocoracoid process may be plesiomorphic characters of *Faxinalipterus*.

Another interesting character of *Faxinalipterus* is the unfused and unreduced fibula. This character is probably the most significant primitive character of *Faxinalipterus*, which neatly differentiates it from the rest of the pterosaurs. However, some exceptions have been recorded in *Dorygnathus* and

Campylognathoides in which "... the fibula ... is developed the full length." (Wellnhofer, 1991, p. 56). In the Late Triassic *Eudimorphodon, Peteinosaurus* and *Preondactylus* (Wild, 1978, 1983), the tibia and fibula are always fused in the proximal area and very reduced at the distal portion. In *Faxinalipterus* the situation is quite different, with an incipient ankylosis proximally but without reduction of length or thickness at the distal portion. Obviously these characters are primitive.

Faxinalipterus was part of a tetrapod association formed by brasilodontid cynodonts (Bonaparte et al., 2003, 2005), tritheledontid cynodonts (Bonaparte et al., 2001; Soares, 2004; Martinelli et al., 2005), the kannemeyerid dicynodont *Jachaleria* (Araújo and Gonzaga, 1980); the clevosaurid sphenodont *Clevosaurus brasiliensis* (Bonaparte and Sues, 2006), the procolophonid parareptilian *Soturnia* (Cisneros and Schultz, 2003), the basal saurischian *Guaibasaurus* (Bonaparte et al., 1999), and a lepidosauriomorph indet. (Bonaparte et al. in preparation). For the first time the association of Pterosauria is recorded with non-mammalian cynodonts and dicynodonts.

As far as it is possible to interpret the meagre fossil material of *Faxinalipterus*, this Brazilian pterosaur along with *Eudimorphodon cromptonellus* may be the most primitive members of the Pterosauria. Because *Faxinalipterus* was recovered from sediments of a continental basin and *E. cromptonellus* from non-marine beds, it is possible that the earliest chapter of pterosaur evolution took place in continental environments, or at least in both continental and littoral marine.

The morphology of the maxilla suggests that the skull was lightly built, with large preorbital and narial openings.

Finally, it is important to note here that the coracoid, humerus and tibia and fibula in *Faxinalipterus* are far more derived than in any other Triassic archosaur such as crocodylomorphs, dinosauromorphs and basal theropods like *Zupaysaurus*. The more robust condition of the humerus compared with that of the femur supports this assumption.

Acknowledgments The authors are thankful to Prof. Kevin Padian, for critical reading of the manuscript and to Pablo Chiarelli, for technical assistance.

References

Araújo, DC, Gonzaga, TD (1980) Uma nova espécie de *Jachaleria* (Therapsida, Dicynodontia) do Triássico do Brasil. 2° Congr Argentino Paleontol Bioestrat 1° Congres Latinoamericano Paleontol, Actas, 2:159–174.

Bonaparte, JF (1973) Edades/Reptil para el Triásico de Argentina y Brasil. Quinto Congres Geol Argentino, Assoc Geol Argentina, 3:93–120.

Bonaparte, JF, Ferigolo, J, Ribeiro, AM (1999) A new early Late Triassic saurischian dinosaur from the Rio Grande do Sul State, Brazil. Proceedings of the 2nd Dinosaur Symposium on National Science Museum Monographs, Tokyo, 15:89–109.

Bonaparte, JF, Ferigolo, J, Ribeiro, AM (2001) A primitive Late Triassic "ictidosaur" from Rio Grande do Sul, Brazil. Palaeontology, 44:623–635.

Bonaparte, JF, Martinelli, AG, Schultz, CL, Rubert, R (2003) The sister group of mammals: small cynodonts from the Late Triassic of southern Brazil. Rev Brasileira Paleontol, 5:5–27.

Bonaparte, JF, Martinelli, AG, Schultz, CL (2005) New information on *Brasilodon* and *Brasilitherium* (Cynodontia, Probainognathia) from the Late Triassic of southern Brazil. Rev Brasileira Paleontol, 8:25–46.

Bonaparte, JF, Sues, H-D (2006) A new species of *Clevosaurus* (Lepidosauria: Rhynchocephalia) from the Upper Triassic of Rio Grande do Sul, Brazil. Palaeontology, 49:917–923.

Cisneros, JC, Schultz, CL (2003) *Soturnia caliodon* n. g. n. sp., a procolophonid reptile from the Upper Triassic of Southern Brazil. N J Geol Palaontol Abh, 227:365–380.

von Huene, F (1914) Beitraege zur Geschichte der Archosaurier. A. Beitraege zur Kenntnis und Beurteilung der Pseudosuchier. 1. Neue Beitraege zur Kenntnis von *Scleromochlus taylori* A.S. Woodward. Geol Palaeontol Abh, 13:1–53.

von Huene, F (1956) Palaeontologie und Phylogenie der niederen Tetrapoden. G. Fisher, Berlin, Germany.

Jenkins, FA, Shubin, NH, Gatesy, SM, Padian, K (2001) A diminutive pterosaur (Pterosauria: Eudimorphodontidae) from the Greenland Triassic. Bull Mus Comp Zool, 156:151–170.

Martinelli, AG, Bonaparte, JF, Schultz, CL, Rubert, R (2005) A new tritheledontid (Therapsida, Eucynodontia) from the Late Triassic of Rio Grande do Sul (Brazil) and its phylogenetic relationships among carnivorous non-mammalian cynodonts. Ameghiniana, 42:191–208.

Padian, K (1983) Osteology and functional morphology of *Dimorphodon macronix* (Buckland), (Pterosauria, Rhamphorhynchoidea): based on new material in the Yale Peabody Museum. Postilla, 189:1–44.

Padian, K (1984) The origin of pterosaurs. In: Reif, W-E, Westphal, F (eds) Proceedings of the 3rd Symposium on Mesozoic Terrestrial Ecosystems, short papers. Attempto, Tübingen, Germany.

Padian, K, Wild, R (1992) Studies on Liassic pterosaurs, I. the holotype and the referred specimens of the Liassic pterosaur *Dorygnathus bauthensis* (Theodori) in the Petrtefaktensammlung Bauz, northern Bavaria. Palaeontographica, A225:59–77.

Rubert, R, Schultz, CL (2004) Um novo horizonte de correlaçáo para o Triássico Superior do Rio Grande do Sul. Pesquisas em Geociencias, 31:71–88.

Soares, MB (2004) Novos materiaisde Riograndia guaibensis (Cynodontia, Tritheledontidae) do Triássico Superior do Rio Grande do Sul, Brasil: análise osteológica e implicações filogenéti-cos. Curso de Pos-Graduação em Geociências, Universidade General de Rio Grande do Sul, Porto Alegre, RS, 347p.

van der Meulen, MCH, Padian, K (1992) Taxon-specific and functional adaptation characteristics of pterosaur bones. Proceedings of the 38th Annual Meeting of the Transactions Orthopedic Research Society, 17:537.

Wellnhofer, P (1974) *Campylognathoides liasicus* (Quenstedt), an Upper Liassic pterosaur from Holzmaden, The Pittsburgh specimen. Ann Carnegie Mus Nat Hist, 45:5–34.

Wellnhofer, P (1991) The illustrated encyclopedia of pterosaurs. Salamander Books, London, UK.

Wild, R (1971) *Dorygnathus mistelgauensis* n. sp., ein neuer Flugsaurier aus dem Lias Epsilon von Mistelgau (Fränkischer Jura). Geol Bull Nordost-Bayern, 21:178–195.

Wild, R (1978) Die Flugsaurier (Reptilia, Pterosauria) aus der oberen Trias von Cene bei Bergamo, Italien. Bollettinoi della Societá Paleontol Italiana, 17:176–256.

Wild, R (1984) A new pterosaur (Reptilia, Pterosauria) from the Upper Triassic (Norian) of Friuli, Italy. Gortania-Atti del Museo Friulano di Storia Naturale, 5:45–62.

Zambelli, R (1973) *Eudimorphodon rauzii* gen. nov. sp. nov., un pterosaurio Triassico. Rendesvous Sci Inst Lombardo, B107:27–32.

Chapter 5
Bone Histology of a Kannemeyeriid Dicynodont *Wadiasaurus*: Palaeobiological Implications

Sanghamitra Ray, Saswati Bandyopadhyay, and Ravi Appana

5.1 Introduction

Dicynodonts formed a major component of the Permian and Triassic terrestrial ecosystem, and are widely used for global correlation and non-marine biochronology. One of the major groups of the Triassic dicynodonts were the large kannemeyeriids with a body size ranging between 1 and 3 m in length (Walter, 1988), elongated and pointed snout, anteriorly placed jaw articulation, and oblique occiput (Bandyopadhyay, 1988). Several kannemeyeriid genera are currently recognized from all over the world, including the genus *Wadiasaurus* recovered from the Middle Triassic Yerrapalli Formation of the Pranhita-Godavari basin of India (Fig. 5.1) (RoyChowdhury, 1970). The taxonomy, morphological description and functional anatomy of *Wadiasaurus* (*W. indicus*) are well established (RoyChowdhury, 1970; Bandyopadhyay, 1988; Ray, 2006). It was a large dicynodont with a skull length of about 400 mm. The characteristic features of *Wadiasaurus* include a low nasal ridge, vertically projected maxillary flanges, wide interpterygoid vacuity (RoyChowdhury, 1970; Bandyopadhyay, 1988), narrow intertemporal region in comparison to a wide interorbital region (IO/IT = 4.7), a tall, narrow scapular blade with a short acromion process and a prominent olecranon process. A functional anatomical study by Ray (2006) shows that *Wadiasaurus* had a near sagittal articulation of the humerus, an efficient forearm movement (extension and flexion), a reduced/restricted lateral undulation of the vertebral column, and a semi-erect/adducted stance of the hindlimbs.

Wadiasaurus is represented by a considerable collection of well-preserved fossil material. At least 700 cranial and postcranial elements amounting to more than 23 individuals of varying age group are known from a single bone bed (Bandyopadhyay, 1988). These specimens had suffered almost no transportation and taphonomic analysis indicated mass mortality due to some catastrophic event

S. Ray (✉)
Department of Geology and Geophysics, Indian Institute of Technology, Kharagpur 721302, India
e-mail: sray@gg.iitkgp.ernet.in

S. Bandyopadhyay (ed.), *New Aspects of Mesozoic Biodiversity*,
Lecture Notes in Earth Sciences 132, DOI 10.1007/978-3-642-10311-7_5,
© Springer-Verlag Berlin Heidelberg 2010

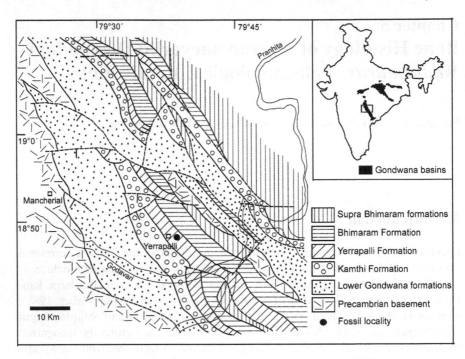

Fig. 5.1 Geological map of the fossil locality (modified after Kutty et al., 1987)

(Bandyopadhyay et al., 2002). However, there is almost no information on their overall life history strategies/palaeobiology.

Bone microstructure of fossil animals provides valuable biological information. Although organic components of the bones (such as the osteocytes, lymph and blood vessels, collagen fibres) get destroyed during fossilization their structural organization remains intact. This allows the bone microstructure to be compared with that of living animals leading to the interpretation of various aspects such as growth pattern, ontogenetic stage and lifestyle adaptations. Osteohistology has contributed significantly to the understanding of various extinct animals such as the non-avian dinosaurs (de Ricqlès, 1972, 1976; Reid, 1984; Chinsamy, 1993; Varricchio, 1993; Curry, 1999; Horner et al., 2000, 2001), other archosaurs (Padian et al., 2004; Cubo et al., 2005; Chinsamy and Elzanowski, 2001), and therapsids (Botha and Chinsamy, 2000, 2004, 2005; Botha, 2003; Ray and Chinsamy, 2004; Ray et al., 2004, 2005; Chinsamy and Hurum, 2006). Hence, bone histology is used in this study to assess the palaeobiology of the Indian dicynodont *Wadiasaurus*. The current study provides a comprehensive histological analysis of *Wadiasaurus* based on several skeletal elements. It elucidates the growth of *Wadiasaurus* at different ontogenetic stages and discusses its lifestyle adaptations.

Institutional abbreviations ISIR, Indian Statistical Institute, Kolkata, India; R refers to the reptilian collection.

5.2 Material and Methods

Various limb bones (humerus, femur, radius, ulna, fibula) along with several dorsal ribs and scapulae were used for the study (Table 5.1). These specimens are all positively identified as *W. indicus*, and are housed in the Geology Museum, Indian Statistical Institute, Kolkata. The material was collected from a monospecific locality of the Middle Triassic Yerrapalli Formation, Pranhita-Godavari basin (Fig. 5.1).

All the specimens were photographed, morphological variations were noted and standard measurements were recorded, using a Mitutoyo digimatic calliper, which has the precision of 0.01 mm. The humerus (ISIR 175/12) is one of the largest elements in the collection and is considered as belonging to an adult animal. Based on the full skeletal reconstruction of *Wadiasaurus* (Ray, 2006), ISIR 175/12 is standardized to compare the dimensions of other skeletal elements and the examined elements are assigned to different size classes (Table 5.1). These size classes are based on the relative size, gross skeletal morphology and bone microstructure as suggested by Ray and Chinsamy (2004).

Wherever possible several serial transverse sections from the same element were processed following the techniques outlined by Chinsamy and Raath (1992). The

Table 5.1 Size classes with bone dimensions (in mm) and calculated relative bone wall thickness (RBT), optimal k values (in parentheses) of the available specimens of *Wadiasaurus*

	Museum registration no.	Elements	Length (mm)	Percentage of adult	Size classes	RBT% (k)
1	ISIR175/419	Ulna	71.45	17.6	Group A	–
2	ISIR175/420	Radius	82.5	20.31	(<30% adult	–
3	ISIR175/421	Fibula	141	24.3	size)	25.5 (0.41)
4	ISIR175/418	Humerus	174	30		–
5	ISIR175/422	Fibula	176.6	30.43	Group B	–
6	ISIR175/95	Femur	265.6	36.61	(>30%, <60%	28.5 (0.33)
7	ISIR175/75	Ulna	154.5	38.03	adult size)	26.6 (0.45)
8	ISIR175/62	Radius	156	38.4		20.8 (0.5)
9	ISIR175/68	Femur	283.3	39.05		31.7 (0.37)
10	ISIR175/11	Femur	312	43		–
11	ISIR175/174	Tibia	402	69.3	Group C	8.05 (0.63)
12	ISIR175/81	Humerus	553.5	95.4	(>60% adult	–
13	ISIR175/12	Humerus	580.3	100	size)	–
14	ISIR175/158	Scapula	–			–
15	ISIR175/37	Scapula	–			–
16	ISIR175/426	Scapula	50*			15 (0.71)
17	ISIR175/424	Rib	21*			–
18	ISIR175/427	Rib	7*			25.8 (0.48)
19	ISIR175/428	Rib	10*			18.9 (0.62)
20	ISIR175/429	Rib	12*			26.5 (0.47)
21	ISIR175/430	Rib	17*			–

Asterisk indicates measurement of the width/diameter.

relative bone wall or cortical thickness (RBT) was measured across the transverse sections and is expressed as a percentage of the diameter (Bühler, 1986; Ray and Chinsamy, 2004). The optimal k value (ratio of the internal diameter to external diameter) as suggested by Currey and Alexander (1985) is also calculated for the available specimens. Since in most of the elements, there is a distinct differentiation between the compact cortex and relatively spongiosa medulla, measurements were taken along two perpendicular directions at 2.5X magnification. Histological terminology and definitions generally followed that of Francillon-Vieillot et al. (1990), de Ricqlès et al. (1991), and Reid (1996).

5.3 Histological Description

The transverse sections of all the bones examined, irrespective of the skeletal element, show a well differentiated centrally located medullary cavity/spongiosa surrounded by an outer relatively compact cortex. Most of the vascular channels show centripetal deposition of osteonal bone and form primary osteons. These primary osteons vary in orientation, and have radial and circumferential anastomoses. These primary osteons are embedded in a woven fibred bone matrix and form fibrolamellar bone tissue. Apart from this basic feature, each element has distinctive features, which are discussed in the following section, according to the various size classes.

5.3.1 Size Class A – Juveniles (<30% Adult Size)

Several limb bones were examined to reveal preponderance of well-vascularized fibrolamellar bone tissue in the cortex (Fig. 5.2). Although the humerus (ISIR 175/418) is diagenetically altered, a reticular pattern of the primary osteons is

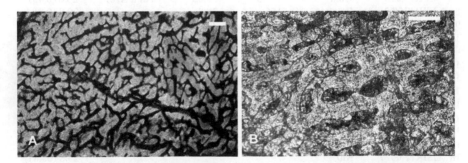

Fig. 5.2 Size class A. Transverse sections of the (**a**) ISIR 175/418, humerus showing well vascularized fibrolamellar bone with reticular pattern in the mid-cortical region; (**b**) ISIR 175/419, ulna showing laminar pattern of the primary osteons in the deeper cortex. *Scale bars* represent 300 μm

evident (Fig. 5.2a). Similar dominance is also observed in the other limb bones of this size class (e.g., ulna, ISIR 175/419, Fig. 5.2b). In the ulna (ISIR 175/419) the orientation of the primary osteons varies from place to place, often having a sub-laminar pattern. Other characteristic features of these bones include a large medullary spongiosa, irregular subperiosteal periphery, absence of growth marks [annuli and lines of arrested growth (LAGs)] and a high cortical porosity.

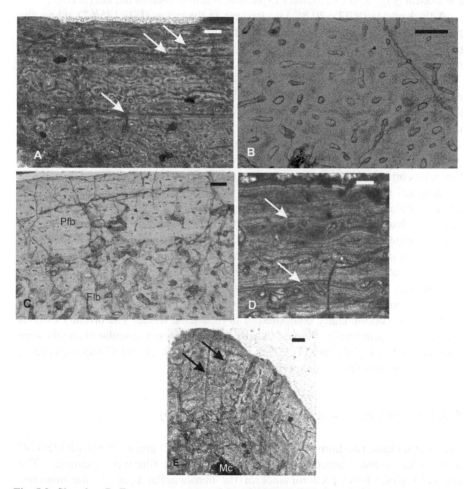

Fig. 5.3 Size class B. Transverse sections of the (**a**) ISIR 175/75, ulna showing zonal plexiform fibrolamellar bone and multiple LAGs (*arrows*); (**b**) ISIR 175/95, femur showing longitudinally oriented primary osteons in the mid-cortical region; (**c**) ISIR 175/68, femur showing a narrow peripheral cortex with parallel fibred bone (Pfb) and a deeper cortex with fibrolamellar bone tissue (Flb); (**d**) ISIR 175/11, femur showing zonal parallel fibred bone and two prominent LAGs (*arrows*); (**e**) ISIR 175/422, fibula showing profuse radially oriented vascular channels (*arrows*). *Scale bars* represent 300 μm

5.3.2 Size Class B – Sub-adult (30–60% Adult Size)

Examination of the various limb bones in this size class reveals a dominance of fibrolamellar bone in cortex. The ulna (ISIR 175/75) has a thick cortex (RBT% = 26.6, Table 5.1) surrounding a distinct medullary cavity. The highly vascularized fibrolamellar bone is zonal and shows a dense laminar-plexiform arrangement (Fig. 5.3a). The ulna is characterized by a high prevalence of radially oriented vascular channels at the medial and lateral edges. Three prominent LAGs are evident (Fig. 5.3a). Medullary expansion is noted towards the lateral edge.

A thick cortex (RBT=20.8%) is also seen in the radius (ISIR175/62), where the primary bone tissue is fibrolamellar. The radius is characterized by a high cortical porosity and an extensive occurrence of resorption cavities in the perimedullary region. Growth marks are absent.

The transverse sections of three femora ranging from 36.6 to 43% adult size were examined to reveal very thick cortices (Table 5.1). In the smaller femur (ISIR175/95) the cortex comprises fibrolamellar bone where the primary osteons are longitudinally oriented, discrete and isolated (Fig. 5.3b). However, the larger femora (ISIR 175/68 and ISIR175/11) shows a more regular, often laminar arrangement of the primary osteons with the elongated osteocyte lacunae aligned in parallel lines in the outer cortex suggesting that the bone tissue is parallel-fibred (Fig. 5.3c, d). The metaphyseal section of the largest femur ISIR175/11 reveals profuse radially oriented vascular channels running from the midcortex to outer cortex in the region of the trochanter major. Bone microstructure of all the femora examined reveals multiple growth marks. For example, a LAG is present in the outer cortex of the smaller femur, ISIR175/95 whereas two thick annuli composed of avascular lamellated bone, and a LAG can be seen in the outer cortex of ISIR175/68. Two LAGs are also seen in the largest femur ISIR175/11 (Fig. 5.3d).

As in other limb bones the cortex of the fibula is thick (for ISIR175/422, RBT = 25.5%) and predominantly composed of fibrolamellar tissue. The primary osteons are mainly longitudinally oriented though radially oriented vascular channels were also seen at the medial edge (Fig. 5.3e). Growth marks are absent. Compacted coarse cancellous bone is seen in the perimedullary region.

5.3.3 Size Class C – Adults (>60% Adult Size)

In this size class, two humeri (ISIR175/81, ISIR175/12) and a tibia (ISIR175/174) along with several scapulae and fragments of dorsal ribs were examined. The cortical region shows predominance of the fibrolamellar bone. In the humerus, ISIR175/81, the primary osteons are longitudinally oriented, discrete and isolate (Fig. 5.4a), though at places these are arranged in a laminar network, and the osteocyte lacunae are flattened and aligned in a particular direction suggesting a peripheral zone of parallel fibered bone (Fig. 5.4b).

Although the tibia (ISIR175/174) is poorly preserved, it shows a large medullary spongiosa surrounded by a narrow cortex (RBT = 8.05%, Table 5.1). The primary

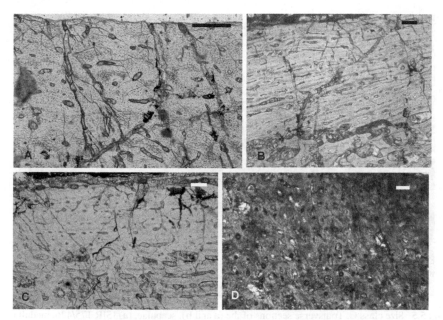

Fig. 5.4 Size Class C. Transverse sections of the (**a–c**), humeri showing (**a** and **b**), ISIR 175/81; (**a**) fibrolamellar bone with discrete primary osteons in the deeper cortex; (**b**) parallel alignment of vascular channels in the outer cortex; (**c**) ISIR 175/12, resorption cavities in the deeper cortex; (**d**) ISIR 175/174, tibia showing profuse primary osteons at the cnemial crest. *Scale bars* represent 300 μm

cortical bone is fibrolamellar. Growth marks are absent. The tibia is characterized by a high porosity and profuse longitudinally and radially oriented vascular channels at the region of the cnemial crest (Fig. 5.4d).

The transverse sections of the three scapular blades (ISIR175/158, ISIR175/426 and ISIR 175/37) show central cancellous regions containing bony trabeculae and bordered on either side by narrow compact cortices, which result in diploes (*sensu* Francillon-Vieillot et al., 1990). In ISIR 175/426, the cortical bone (RBT = 15%) is fibrolamellar while extensive secondary reconstruction is seen in the perimedullary region (Fig. 5.5a). On the other hand, in ISIR175/158 the peripheral region of the cortex shows laminar pattern of the primary osteons and the bone matrix changes to parallel fibered (Fig. 5.5b). Similarly in the other scapula (ISIR175/37), the primary osteons are organized in a laminar network mainly in the outer cortex. A LAG is possibly present in ISIR175/426 (Fig. 5.5a). All the scapulae examined are characterized by numerous erosionally enlarged resorption cavities in the perimedullary region, which is followed by a zone of endosteal bone.

In all the ribs examined, except ISIR175/430, there is a clear distinction between the cortex (RBT ranging between 18.9 and 26.5%) and the medullary cavity (Fig. 5.5c). In ISIR175/427, ISIR175/428 and ISIR 175/429, the cortex comprises extensive secondary reconstruction in the form of large resorption cavities that

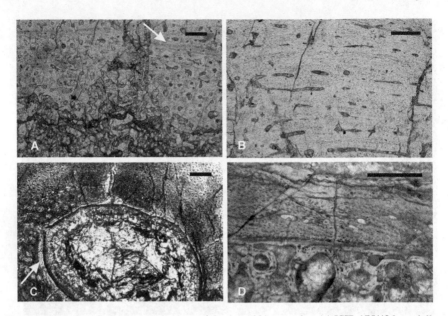

Fig. 5.5 Size class C. Transverse sections of the (**a** and **b**), scapulae, (**a**) ISIR 175/426, medullary spongiosa and resorption cavities in the deeper cortex. A LAG is shown by an *arrow*; (**b**) ISIR 175/37, showing parallel fibred bone in the cortical region; (**c** and **d**) dorsal ribs, (**c**) ISIR 175/427, showing longitudinally oriented vascular channels in the cortex and a narrow inner circumferential lamellae (*arrow*) surrounding the medullary cavity; (**d**) ISIR 175/429, a narrow compact outer cortex and extensive resorption cavities in the mid- to inner cortical region. *Scale bars* represent 300 μm

almost reaches the outer cortex (Fig. 5.5d). There is a distinct line separating the zone of resorption cavities from the compact cortex. The latter is very narrow in these ribs and comprises of almost avascular parallel-fibred/lamellar bone (Fig. 5.5d). A thin layer of internal circumferential lamellae surrounds the medullary cavity in ISIR175/429 (Fig. 5.5c).

5.4 Discussion

5.4.1 Ontogenetic Growth Pattern

The bone microstructure of the different skeletal elements of *Wadiasaurus* is characterized by the preponderance of the highly vascularized fibrolamellar bone tissue, which constitutes the greater part of the compact cortex. The arrangement of the primary osteons varied from longitudinal and discrete to laminar-subplexiform-reticular. Such patterns may vary depending on the skeletal element, and even locally within the same transverse section. Fibrolamellar bone tissue is considered to indicate rapid osteogenesis and hence an overall fast growth (Amprino, 1947; de

Margerie et al., 2002). Fibrolamellar tissue has also been identified in other dicynodonts (de Ricqlés, 1972; Chinsamy and Rubidge, 1993) such as *Diictodon* (Ray and Chinsamy, 2004), *Oudenodon* (Botha, 2003), *Lystrosaurus* (Ray et al., 2005) and in the non-mammalian therapsids in general (de Ricqlés, 1969; Botha and Chinsamy, 2000, 2004, 2005; Ray et al., 2004).

Although the overall bone tissue generally agrees with that of other dicynodonts, *Wadiasaurus* is characterized by the presence of peripheral parallel-fibred bone tissue, especially in the size classes B and C. Such high prevalence of peripheral parallel fibred bone has not been noted in any other dicynodont except *Oudenodon* (Botha, 2003).

Another notable feature of *Wadiasaurus* bone microstructure is the dominance of laminar network of the primary osteons in most of the skeletal elements examined. Although relationship between the primary bone tissue type (especially based on arrangement of the primary osteons) and bone growth rate is unclear and controversial, a recent study on the king penguin chick by de Margerie et al. (2004) have shown that comparatively laminar fibrolamellar bone had moderate growth rate whereas radially oriented vascular channels in fibrolamellar bone grows faster. In accordance, it may be suggested that within the fast growth strategy, *Wadiasaurus* growth became moderate in the later stages, and eventually slowed down considerably as revealed by the peripheral parallel fibred bone.

Although it was not possible to examine the entire size range because of the paucity of the specimens available for examination, it is evident that *Wadiasaurus* had three distinct ontogenetic stages based on bone microstructure, which is summarized in Table 5.2 and schematically represented in Fig. 5.6. The first or juvenile stage (Growth stage I) was characterized by a well-vascularized fibrolamellar bone in the cortex, a high cortical porosity and the reticular-plexiform arrangement of the primary osteons (Table 5.2). Growth marks were absent. This type of primary tissue and absence of growth marks suggest that growth was fast and continuous during the juvenile stage. Other features include onset of the secondary reconstruction and absence of the endosteal bone deposition. This uninterrupted rapid growth continued upto 30% of adult size. After attaining 30% of adult size, growth became periodically interrupted as evident from the presence of LAGs. In addition, this stage shows the onset of endosteal bone deposition. This intermediate or sub-adult stage (Growth stage II) possibly continued till the individual attained 60% of adult size.

In the last or adult stage (>60% adult size), the bone microstructure was characterized by a peripheral zone of parallel fibred bone, decrease in vascularity towards the periosteal periphery, more organized arrangements of the primary osteons (usually laminar fibrolamellar), extensive secondary reconstruction and endosteal bone deposition and varied incidence of growth lines. Most of the dorsal ribs examined (ISIR175/427–ISIR175/430) had an even margin of the medullary cavity and a thin layer of endosteal lamellated bone tissue (inner circumferential lamellae) surrounding it, suggesting near completion of the medullary expansion (Reid, 1993). An interesting feature noted in all the dorsal ribs examined is the extensive secondary reconstruction in the form of large resorption cavities almost reaching the outer

Table 5.2 Characteristic features of the three growth stages of *Wadiasaurus*

Growth stage I: size class A – Juveniles (individuals <30% adult size)
  Primary fibrolamellar bone in the cortex
  Plexiform-reticular arrangement of the primary osteons
  High cortical porosity
  Irregular periosteal periphery
  No growth marks
  Numerous resorption cavities in the perimedullary region
  No endosteal bone deposition
Growth stage II: size class B – subadults (30–60% adult size)
  Primary fibrolamellar bone in the cortex
  Longitudinally oriented, discrete and isolate primary osteons, often forming
  laminar-plexiform pattern
  Multiple annuli and LAGs present
  Extensive secondary reconstruction
  Onset of endosteal bone deposition
Growth stage III: size class C – adults (>60% adult size)
  Fibrolamellar bone in the deeper cortex
  Peripheral parallel fibred/lamellar bone
  Longitudinally oriented, discrete and isolate primary osteons, often forming
  laminar pattern
  Vascularity decreases towards periosteal periphery
  LAGs present
  Extensive secondary reconstruction
  Endosteal bone deposition – inner circumferential lamellae seen in ribs

---

cortex (Fig. 5.5d) and giving it an apparent appearance of a cancellous bone. This sort of bone microstructure of the ribs has not yet been observed in other dicynodonts. One possible explanation for this feature could be haemopoesis. Since the ribs are non-weight bearing elements, they usually act as blood cell synthesizing elements and are the regions of high metabolic activity and hence growth (Ten Cate, 1994). However, further study is required to ascertain this conclusion.

The overall bone microstructure in this growth stage suggests an initial fast growth followed by a decrease in active bone growth rate and permanent slowing down of overall growth. The transition from the fast growing fibrolamellar bone to the slow growing parallel fibred bone possibly represented the onset of sexual maturity (Reid, 1996; Padian, 1997; Botha, 2003). However, absence of peripheral rest lines suggests that appositional growth had not ceased (but slowed down) in the last ontogenetic stage. Since body growth rate has been related to primary bone growth rate (Amprino, 1947) and given the fact that some of the largest (presumably the oldest) specimens in the collection were examined, it is suggested that *Wadiasaurus* had an indeterminate growth strategy (*sensu* Chinsamy and Hurum, 2006; Foote and Miller, 2007). Such a growth strategy is more in common with that of the amphibians and extant sauropsids, where growth continues even after attaining adult size and contrasts with that of the mammals and birds (Chinsamy and Dodson, 1995; Curry, 1999). These latter groups stopped growing after attaining a maximum size

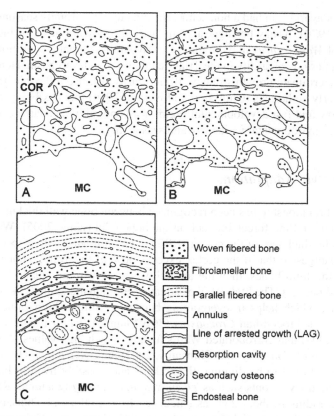

Fig. 5.6 Schematic representation of bone microstructure in the three ontogenetic stages of *Wadiasaurus*. (**a**) Juvenile stage; (**b**) sub-adult stage; (**c**) adult stage. Bone surface is towards the *top* of the page

and the avascular peripheral bone is often marked by multiple growth marks (or rest lines) to show that growth has reached a plateau.

The presence of annuli and LAGs in the different skeletal elements of *Wadiasaurus* reflects environmental stress as seen in the extant crocodilians (Hutton, 1986). Similar LAGs have been also observed in some rodents during hibernation (Klevezal, 1996) whereas Chinsamy et al. (1998) have shown that Arctic polar bears develop LAGs during adverse climatic conditions. Starck and Chinsamy (2002) have suggested that the variable incidence of annuli and LAGs among several dinosaurs and extant vertebrates reflects their variable response to environment. These growth marks suggest high degree of developmental plasticity (*sensu* Smith-Gill, 1983) in *Wadiasaurus*, which is the ability to respond to changes in the environment by evoking different developmental regimes. This ability to stop growth and development during adverse environmental conditions has also been observed in other nonmammalian therapsids, and is considered as a plesiomorphic condition for the mammalian lineage (Ray et al., 2004). In the Triassic period, the

Pranhita-Godavari basin had a hot, semi-arid climate with strongly seasonal rainfall (Robinson, 1970). It may be suggested that periodic slowing down or interruption in growth of *Wadiasaurus* may be due to unfavourable climatic conditions, possibly the long periods of aridity in the Triassic Pranhita-Godavari basin. Hence, along with an indeterminate growth strategy, *Wadiasaurus* also had a flexible growth. As in case of early eutherians (Chinsamy and Hurum, 2006) this flexible growth strategy with slow and fast developmental regimes was possibly an adaptive advantage to overcome environmental variability and seasonality.

5.4.2 Lifestyle Adaptations

A functional relationship has been recognized between bone wall thickness, habitat of the animal, and the forces that act on the bone (Chinsamy, 2005). Wall (1983) considered the thickness of the cortex (RBT) to be correlated with a specific mode of life. He suggested that if the cortex exceeds 30% of the average bone diameter in most of the limb bones, then the animal was at least semi-aquatic (e.g. hippo, manatee and beaver). The high RBT of aquatic/semi-aquatic animals suggests high bone density, which helps in overcoming buoyancy. Similar high values of RBT were seen in the crocodilian *Crocodylus* (36.3%) and *Alligator* (36.45%, Ray and Chinsamy, 2004) that corroborated Wall's findings. Such high values of RBT were also seen in *Diictodon* (Ray and Chinsamy, 2004) and *Oudenodon* (Botha, 2003), which were inferred to be digging dicynodonts and possibly lived in burrows. In non-mammalian cynodonts such as *Thrinaxodon*, a burrowing animal, RBT% of a nearly adult radius is about 28% as calculated from Botha and Chinsamy (2005) whereas that of the fossorial *Trirachodon* is 30% (Botha and Chinsamy, 2004). Hence it appears that a burrowing or fossorial life style also leads to thick bone walls. Variable cortical thickness at the midshaft region of the long bones is also seen in the Mesozoic mammals. For example, femoral RBT% of *Morganucodon* is about 19%, that of the extinct multituberculate *Nemegtbaatar* is 26% and that of extant eutherian *Heterocephalus* varies from is 28.6 to 31% (as calculated from Chinsamy and Hurum, 2006; Botha and Chinsamy, 2004). It is evident from these findings that not only the mode of life but differing functional constraints of the limb bones could result in differing RBT in the animals.

In addition, bone microstructure may be used for interpreting life style adaptations of fossil animals (Chinsamy, 2005). Although most limb bones of the Early Triassic dicynodont *Lystrosaurus* show a cancellous appearance because of extensive resorption cavities, the dorsal ribs are characterized by high cortical thickness, which is about 35.17% after attaining 76.84% adult size (Ray et al., 2005). Such thick dorsal ribs of *Lystrosaurus*, which was also noted by Retallack et al. (2003), contrasted with that of other dicynodonts (Ray et al., 2005) and had been attributed to the semi-aquatic lifestyle of *Lystrosaurus* (Ray et al., 2005). Moreover, certain features in the gross skeletal morphology of *Lystrosaurus* such as a wide scapular blade and paddle-like antebrachium suggests that *Lystrosaurus* probably had a combined semi-aquatic and burrowing habit (Ray et al., 2005).

In *Wadiasaurus* the RBT of the adult limb bones is <30% (except for the femur, Table 5.1). Figure 5.7a gives a comparison of the femoral RBT of *Wadiasaurus* with the limb bones of other available dicynodonts, the cynodonts (*Thrinaxodon, Trirachodon, Morganucodon, Nemegtbaatar* and *Heterocephalus*), extant semi-aquatic animals and terrestrial megaherbivores. It may be mentioned here that only the midshaft regions of the long bones have been examined. The RBT of those dicynodonts inferred to have a burrowing life style is much higher (>36%), than that of *Wadiasaurus* except for *Lystrosaurus*. The variation seen in the latter taxon is probably a result of its combined semi-aquatic and burrowing life style.

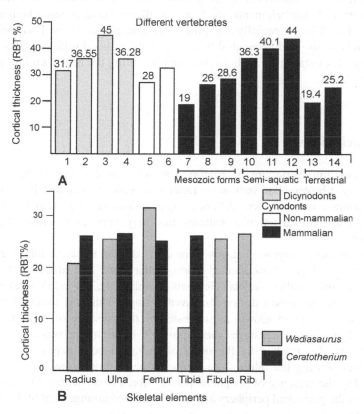

Fig. 5.7 (a) Comparative cortical thicknesses of different dicynodonts and extant animals exhibiting different lifestyle adaptations. RBT% measured at the midshaft regions of the long bones. All long bones measured are femora except for those given in parentheses. Legends: 1, *Wadiasaurus*; 2, *Lystrosaurus* (dorsal rib); 3, *Oudenodon*; 4, *Diictodon*; 5, *Thrinaxodon* (radius); 6, *Trirachodon*; 7, *Morganucodon*; 8, *Nemegtbaatar*; 9, *Heterocephalus*; 10, *Alligator*; 11, *Odobenus* (Walrus); 12, *Hippopotamus*; 13, *Bison*; 14, *Ceratotherium* (White rhino). (b) Comparative cortical thickness of different skeletal elements of adult *Wadiasaurus* and *Ceratotherium* (White rhino). All are measured at the midshaft region. Sources of information: Wall (1983), Botha (2003), Ray and Chinsamy (2004), Botha and Chinsamy (2004, 2005), Ray et al. (2005), and Chinsamy and Hurum (2006)

Along with low to moderate RBT values, there are no morphological adaptations (Ray, 2006) to suggest that *Wadiasaurus* was a semi-aquatic/aquatic animal. In addition, the RBT of the forelimb bones of *Wadiasaurus* (except femur) is comparable with that of the extant megaherbivores such as *Ceratotherium* (Fig. 5.7b), though the femoral thickness is much higher. The calculated k-values of the adult limb bones (ranging between 0.3 and 0.6) show that *Wadiasaurus* is comparable with land animals (Currey and Alexander, 1985), where the limb bones were selected for impact loading, and were able to absorb a certain amount of kinetic energy without breaking the bone. Hence, cortical thickness and optimal k-values suggest that *Wadiasaurus* may be considered as a generalized terrestrial animal, possibly a megaherbivore, which is further corroborated by its skeletal design (Ray, 2006). However, compared to other skeletal elements, the femur of *Wadiasaurus* shows a high cortical thickness (Fig. 5.7b), especially along the lateral and medial edges. This could possibly be related to the dorsoventral flattening of the femoral midshaft (Ray, 2006) for enhanced resistance to mediolateral bending during the nearly upright posture of the hindlimbs.

5.5 Conclusions

1. The bone microstructure of *Wadiasaurus* revealed that the cortical bone is primarily fibrolamellar indicating rapid osteogenesis and fast bone growth. The primary osteons show differing organizations ranging from isolated and discrete to laminar-reticular patterns that may vary even within the same section.
2. Three distinct ontogenetic stages were inferred from the bone microstructure. Presence of highly vascularized fibrolamellar bone and absence of growth marks in the smaller skeletal elements examined (<30% adult size) suggest sustained rapid growth during the juvenile stage. This was followed by periodic interruptions in growth as suggested by the presence of multiple growth marks in the sub-adult stage when upto 60% of adult size was attained. This stage was also marked by the onset of endosteal bone deposition. Later in ontogeny, during the adult stage, growth slowed down considerably as suggested by the presence of peripheral parallel fibred bone, decrease in vascularity towards the periosteal periphery and more organized arrangement of the primary osteons.
3. The onset of the slow growth possibly marks sexual maturity. Presence of multiple growth marks in the sub-adult and adult stages suggests a flexible growth strategy whereas the absence of peripheral rest lines in the largest and presumably oldest specimen examined indicates that *Wadiasaurus* had an indeterminate growth strategy, similar to that of present day sauropsids.
4. Bone cross sectional geometry reveals that most of the adult limb bones of *Wadiasaurus* have RBT<30% (except femur) and k values ranging between 0.3 and 0.6, which are comparable with that of the present day terrestrial vertebrates.

Acknowledgments   The authors are grateful to Ms. D. Mukherjee, Department of Geology & Geophysics, Indian Institute of Technology, Kharagpur for technical assistance. The authors thank Prof. A Chinsamy, University of Cape Town, South Africa and Prof. A. Sahni, Panjab University, India for review and constructive criticism. Council for Scientific and Industrial Research, India, Indian Statistical Institute, Kolkata, Indian Institute of Technology, Kharagpur, and Department of Science and Technology, India provided fund and infrastructural support.

References

Amprino, R (1947) La structure du tissu osseux envisagée comme expression de différences dans la vitesse de l'accroissement. Arch de Biol, 58:315–330.

Bandyopadhyay, S (1988) A kannemeyeriid dicynodont from the Middle Triassic Yerrapalli Formation. Phil Trans R Soc Lond, B320:185–233.

Bandyopadhyay, S, RoyChowdhury, TK, Sengupta, DP (2002) Taphonomy of some Gondwana vertebrate assemblages of India. Sed Geol, 147:219–245.

Botha, J (2003) Biological aspects of the Permian dicynodont *Oudenodon* (Therapsida: Dicyno-dontia) deduced from bone histology and cross-sectional geometry. Palaeontol Afr, 39:37–44.

Botha, J, Chinsamy, A (2000) Growth patterns deduced from the histology of the cynodonts *Diademodon* and *Cynognathus*. J Vertbr Paleontol, 20:705–711.

Botha, J, Chinsamy, A (2004) Growth and life habits of the Triassic cynodont *Trirachodon*, inferred from bone histology. Acta Palaentol Polonica, 49:619–627.

Botha, J, Chinsamy, A (2005) Growth patterns of *Thrinaxodon liorhinus*, a non-mammalian cynodont from the Lower Triassic of South Africa. Palaeontol, 48:385–394.

Bühler, P (1986) Das Vogelskellet – hochentwickelter knochen-leichtbau. Arcus, 5:221–228.

Chinsamy, A (1993) Bone histology and growth trajectory of the prosauropod dinosaur *Massospondylus carinatus* Owen. Mod Geol, 18:319–329.

Chinsamy, A (2005) The microstructure of dinosaur bone: deciphering biology with fine-scale techniques. The Johns Hopkins University Press, Baltimore, MD.

Chinsamy, A, Dodson, P (1995) Inside a dinosaur bone. Am Sci, 83:174–180.

Chinsamy, A, Elzanowski, A (2001) Evolution of growth patterns in birds. Nature, 412:402–403.

Chinsamy, A, Hurum, JH (2006) Bone microstructure and growth patterns of early mammals. Acta Palcontol Polonica, 51:325–338.

Chinsamy, A, Raath, MA (1992) Preparation of fossil bone for histological examination. Palaeontol Afr, 29:39–44.

Chinsamy, A, Rich, T, Vickers-Rich, P (1998) Polar dinosaur bone histology. J Vertbr Paleontol, 18:385–390.

Chinsamy, A, Rubidge, BS (1993) Dicynodont (Therapsida) bone histology: phylogenetic and physiological implications. Palaeontol Afr, 30:97–102.

Cubo, J, Ponton, F, Laurin, M, de Margerie, E, Castanet, J (2005) Phylogenetic signals in bone microstructure of sauropsids. Syst Biol, 54:562–574.

Currey, JD, Alexander, RM (1985) The thickness of tubular bones. J Zool, A206:453–468.

Curry, KA (1999) Ontogenetic histology of *Apatosaurus* (Dinosauria: Sauropoda): new insights on growth rates and longevity. J Vertbr Paleontol, 19:654–665.

de Margerie, E, Cubo, J, Castanet, J (2002) Bone typology and growth rate: testing and quantifying 'Amprino's rule' in the mallard (*Anas platyrhynchos*). C R Biol, 325:221–230.

de Margerie, E, Robin, JP, Verrier, D, Cubo, J, Groscolas, R, Castanet, J (2004) Assessing a rela-tionship between bone microstructure and growth rate: a fluorescent labeling study in the king penguin chick (*Aptenodytes patagonicus*). J Exp Biol, 207:869–879.

de Ricqlès, A (1969) Recherches palèohistologiques sur les os long Tètrapodes. II, quelque observations sur la structure des log des Thèriodontes. Annales de Palèontologie (Vertèbres), 55:1–52.

de Ricqlès, A (1972) Recherches paléohistologiques sur les os longs des Tétrapodes. Part III. Titanosuchiens, Dinocephales, Dicynodontes. Ann Paléontol (Vertbr), 58:17–60.

de Ricqlès, A (1976) On bone histology of fossil and living reptiles with comments on its functional and evolutionary significance. In: Bellairs, A d'A, Cox, CB (eds) Morphology and biology of reptiles. Linnean Society Symposium Series 3, London, UK.

de Ricqlès, A, Meunier, FJ, Castanet, J, Francilon-Vieillot, H (1991) Comparative microstructure of bone. In: Hall, BK (ed) Bone 3: bone matrix and bone specific products. CRC Press Inc, Boca Raton, FL.

Foote, M, Miller, AI (2007) Principles of paleontology, 3rd edn. W. H. Freeman and Company, New York, NY.

Francillon-Vieillot, H, de Buffrénil, V, Castanet, J, Gerandie, J, Meunier, FJ, Sire, JY, Zylberberg, LL, Ricqlès, A (1990) Microstructure and mineralization of vertebrate skeletal tissues. In: Carter, JG (ed) Skeletal biomineralization: patterns, process and evolutionary trends 1. Van Nostrand Reinhold, New York, NY.

Horner, JR, Padian, K, de Ricqlès, A (2001) Comparative osteohistology of some embryonic and perinatal archosaurs: developmental and behavioral implications for dinosaurs. Paleobiol, 27:39–58.

Horner, JR, de Ricqlès, A, Padian, K (2000) Long bone histology of the hadrosaurid dinosaur *Maiasaura peeblesorum*: growth dynamics and physiology based on an ontogenetic series of skeletal elements. J Vertbr Paleontol, 20:115–129.

Hutton, JM (1986) Age determination of living Nile crocodiles from the cortical stratification of bone. Copeia, 1986:332–341.

Klevezal, GA (1996) Recording structures of mammals. A A Balkema Publishers, Brookfield, VT.

Kutty, TS, Jain, SL, RoyChowdhury, TK (1987) Gondwana sequence of the northern Pranhita-Godavari valley: its stratigraphy and vertebrate faunas. Palaeobotanist, 36:214–229.

Padian, K (1997) Growth lines. In: Currie, PJ, Padian, K (eds) Encyclopedia of dinosaurs. Academic Press, San Diego, CA.

Padian, K, Horner, JR, de Ricqlès, A (2004) Growth in small dinosaurs and pterosaurs: the evolution of archosaurian growth strategies. J Vertbr Paleontol, 24:555–571.

Ray, S (2006) Functional and evolutionary aspects of the postcranial anatomy of dicynodonts (Synapsida, Therapsida). Palaeontol, 49:1263–1286.

Ray, S, Botha, J, Chinsamy, A (2004) Bone histology and growth patterns of some nonmammalian therapsids. J Vertbr Paleontol, 24:634–648.

Ray, S, Chinsamy, A (2004) *Diictodon feliceps* (Therapsida, Dicynodontia): bone histology, growth and biomechanics. J Vertbr Paleontol, 24:180–194.

Ray, S, Chinsamy, A, Bandyopadhyay, S (2005) *Lystrosaurus murrayi* (Therapsida; Dicyndontia): bone histology, growth and lifestyle adaptations. Palaeontol, 48:1169–1185.

Reid, REH (1984) Primary bone and dinosaurian physiology. Geol Mag, 121:589–598.

Reid, REH (1993) Apparent zonation and slowed late growth in a small Cretaceous theropod. Mod Geol, 18:391–406.

Reid, REH (1996) Bone histology of the Cleveland-Llyod dinosaurs and of dinosaurs in general, Part 1: Introduction to bone tissues. Geol Stud, 41:1–71.

Retallack, GJ, Smith, RMH, Ward, PD (2003) Vertebrate extinction across P-T boundary in Karoo basin, South Africa. Geol Soc Am Bull, 115:1133–1152.

Robinson, PL (1970) The Indian Gondwana formations – a review. Proceedings of the 1st Inter-national Symposium on Gondwana Stratigraphy, IUGS. Buenos Aires, Argentina, pp. 201–268.

RoyChowdhury, T (1970) Two new dicynodonts from the Triassic Yerrapalli Formation of central India. Palaeontol, 13:132–144.

Smith-Gill, SJ (1983) Developmental plasticity: developmental conversion versus phenotypic modulation. Am Zool, 23:47–55.

Starck, JM, Chinsamy, A (2002) Bone microstructure and developmental plasticity in birds and other dinosaurs. J Morphol, 254:232–246.

Ten Cate, AR (1994) Oral histology: development, structure and function, 4th edn. Mosby-Year Book Inc, St. Louis, MO.

Varricchio, DJ (1993) Bone microstructure of the Upper Cretaceous theropod dinosaur *Troodon formosus*. J Vertbr Paleontol, 13:99–104.

Wall, WP (1983) The correlation between high limb bone density and aquatic habits in recent mammals. J Palaeontol, 57:197–207.

Walter, LR (1988) The limb posture of kannemeyeriid dicynodonts: functional and ecological considerations. In: Padian, K (ed) The beginnings of the age of dinosaurs. Cambridge University Press, Cambridge, MA.

Chapter 6
Indian Cretaceous Terrestrial Vertebrates: Cosmopolitanism and Endemism in a Geodynamic Plate Tectonic Framework

A. Sahni

6.1 Introduction

The Indian non-marine stratigraphic record from the Middle-Late Cretaceous to the Lower Eocene is now fairly well known and studied through a remarkable series of palaeontological finds. These range from the oldest global record of micrometer-sized diatoms (Ambwani et al., 2003) through the presence of gigantic dinosaurs (Khosla et al., 2003) to the beginning of the earliest great radiation of mammals at about 52 Ma that took place after the Palaeocene-Eocene Thermal Maximum (PETM) event (Rose et al., 2006, 2009; Kumar, 2001; Sahni et al., 2006; Thewissen et al., 2001). Apart from the evolutionary and palaeobiogeographical implications of this biotic assemblage (Bajpai et al., 2005; Clyde et al., 2003; Gingerich et al., 1997) the data base from the Indian subcontinental plate provides insights into the biotic relationships of other small and large southern hemisphere lithospheric plates which have broken free in the past and are drifting northwards.

In the Indian subcontinent the stratigraphic record of terrestrial faunas and floras with which this paper is concerned extends from peninsular India (mainly Cretaceous in age), through the Palaeocene (mainly floras) to the Lower Eocene represented by the early Tertiary coal measures of India and Pakistan (Bajpai and Thewissen, 2002; Rana et al., 2005; Sahni et al., 2006) and the Middle Eocene (Russell and Gingerich, 1981; Sahni and Jolly, 1993). Good terrestrial sections are also available in the Neogene (Garzanti et al., 1986; Sahni, 1999) but these will not be discussed here.

Relevant to the present review are sections that have yielded terrestrial vertebrates and floras including the Cenomanian-Turonian of the Bagh-Nimar area (Bagh Group) of Madhya Pradesh (Khosla et al., 2003) and the widespread Deccan Trap associated sedimentary sequences (DTASS) which are highly fossiliferous (Khosla and Sahni, 2003; Prasad, 2005; Prasad and Khajuria, 1995, 1996; Prasad and Rage, 1991, 1995, 2004; Prasad and Sahni, 1988). The Lower Eocene sequences have been

A. Sahni (✉)
Centre of Advanced Study in Geology, Panjab University, Chandigarh 160014, India
e-mail: ashok.sahni@gmail.com

S. Bandyopadhyay (ed.), *New Aspects of Mesozoic Biodiversity*,
Lecture Notes in Earth Sciences 132, DOI 10.1007/978-3-642-10311-7_6,
© Springer-Verlag Berlin Heidelberg 2010

discussed in a general geological and chronological framework (Sahni et al., 2006; Serra-Kiel et al., 1998) while the subcontinental Middle Eocene has been widely studied both in Pakistan and India by several workers (Kumar, 2001; Thewissen et al., 2001).

6.2 Bagh and Nimar Biotas: Sharks, Gigantic Sauropods

The separation of the western India margin from Madagascar is a well documented event both on geophysical as well as on palaeontological bases (Bardhan et al., 2002). There is growing evidence that extensional sea floor spreading between the two island subcontinents did not hamper migration of terrestrial biotas until at least the latest Cretaceous. The Nimar Formation which underlies the shark-bearing Bagh Formation (Bagh Group) has yielded gigantic sauropod skeletal elements along with petrified tree trunks in an essentially intertidal, highly bioturbated clastic sequence (Khosla et al., 2003). This facies is similar to the one described for the Bahariya Formation of Egypt (Smith et al., 2001) which is also of the same age, namely Cenomanian-Turonian. The sauropod elements have yet to be taxonomically assigned but are considered close to the Gondwanan form *Saltasaurus* (Khosla et al., 2003).

6.3 Deccan Trap Associated Sedimentary Sequences (DTASS)

The next major terrestrial biota bearing assemblages are from the DTASS which are largely Maastrichtian in age and comprise a volcano-sedimentary suite of rocks found essentially on the peripheral margins of the extensive Deccan Volcanic Province. They also occur wherever the lava pile has been deeply dissected by the drainage as in the case of the classic Jabalpur, (Madhya Pradesh) localities (Sahni et al., 1994). The rationale behind using the acronym DTASS is illustrated in Fig. 6.1. It should be noted that the terms infra- and inter-trappeans have created confusion for over 100 years and should now be discarded. As explained in Fig. 6.1, in a Deccan volcanic context all the sedimentary beds are truly "inter-trappean" in a temporal sense as they post-date the initial volcanic eruptions in a spatial framework. However locally sedimentary beds may lie below the local basal traps suggesting that there may be a temporal connotation to their older stratigraphic position. This may certainly not be the case and it is possible that a so-called "inter-trappean" may in fact be stratigraphically "older" than an "infratrappean" (Fig. 6.1, Section A). The important point for consideration is that it is the position of the snout of the lava flow/flows is what determines whether an "infratrappean" is further divided to become an "intertrappean" or not. In the field it is difficult to find the exact position of the snout of a lava flow as it is a point in space; it is more likely that one will follow the strike line of the flow and therefore it is difficult to see the gradation of an intertrappean into an infratrappean or vice versa. Based on biotic and

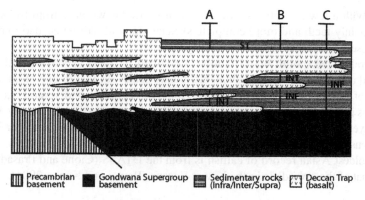

Fig. 6.1 This diagrammatic representation illustrates the interrelationship of the sedimentary beds associated with the Deccan lava flows in a temporal and spatial framework. On a wide regional basis, all sedimentary beds are intertrappeans irrespective of their local position as all studied sequences show the influence of volcanic activity in terms of sedimentary processes and the common presence of biotic elements. Perceived differences are probably related to differing facies and depositional environments. The extension of the lava flow front represented by the moving snout of the lava flow/flows determines whether an "intertrappean" will be further subdivided or not

chronologic criteria alone, the two sets of sedimentary beds cannot be differentiated except possibly on a facies basis (Sahni et al., 1994; Tandon et al., 1995).

In the following section, the known biotic components of the DTASS are discussed in terms of endemism, cosmopolitanism and evolutionary implications within their geodynamic setting.

6.3.1 Fish

Characteristic fish assemblages are recorded from the DTASS. These include about 10–15 families including those based on otoliths (Rana, 1988). They represent largely a coastal plain complex with rays, holosteans and teleosts. The myliobatid ray that has been considered to be the most important in terms of spatial and temporal distribution is *Igdabatis*, first described by Cappetta (1972) from the Mount Igdaman region of Niger in what are considered to be typical marine deposits associated with sharks and other marine forms. *Igdabatis* is now known from several localities in India both in the DTASS context and in more open marine conditions as in Fatehgarh as well as from northern Africa and Spain (Prasad and Cappetta, 1993; Soler-Gijon and Martinez, 1998). This distribution implies an Indian – Eurasian connection even in the latest Cretaceous (Khosla and Sahni, 2003). The taxon appears to be the most dominant part of the fish assemblage mainly because the teeth are robust and have a better chance of preservation in relation to other fish. It should be borne in mind that titanosaurid dinosaurs are also found in both India and the northern Mediterranean Cretaceous localities from where *Igdabatis* has been recovered.

The next most common taxa are osteoglossids and lepisosteids and these are better documented by preserved skulls and other skeletal materials (Sahni and Jolly,

1993). With a few exceptions, other fish genera are known only from isolated elements mainly teeth and more complete skeletal material will be needed to confirm their presence. Pycnodonts are also fairly common.

Several of the otolith based families described by Rana (1988) have also been recorded on the basis of other elements mainly scales. In fact in one of the earliest reports of fish from the Intertrappeans, Hora (1939) mentioned several taxa based on scales including clupeids, which he considered as suggestive of an early Tertiary age. Gayet et al. (1984) studied a new assemblage from Gitti Khadaan near Nagpur and discussed the palaeobiogeographic importance of the fish assemblages.

The oldest Asian record of catfish is from the DTASS (Cione and Prasad, 2002) while that of the gobiids is in the Lower Eocene (Bajpai and Kapur, 2004).

6.3.2 Amphibians

So far the frogs are represented by at least by five families, Leptodactylidae, ? Hylidae, Ranidae, Pelobatidae and the Discoglossidae. The identity of the frogs is a contentious issue as in the case of the pelobatids and the discoglossids, the taxa have been identified only on the basis of isolated limb elements. However, the evidence has been sufficient to suggest that the frogs may have had a Laurasiatic affinity (Prasad and Rage, 1991, 1995, 2004) and this data has been used in part to postulate an early (Late Cretaceous) age for the India-Asia Collision (Jaeger et al., 1989; Prasad et al., 1995).

On the other hand the better preserved frogs from the classically studied Worli Frog Beds of the Mumbai area belong to the Leptodactylidae (Spinar and Hodrova, 1985) which suggests that the frogs are closer to their Gondwanan counterparts. A recent find of a living frog species (*Nasikabatrachus sahyadriensis*) has had great significance for frog systematics and palaeobiogeography (Biju and Bossuyt, 2003), as it demonstrates the past contiguity of the Seychelles island complex which was united with Greater India about 65 Ma. In both these now widely separated areas there is the common presence of closely related primitive frogs (Nasikabatrachidae from the Western Ghat region and the Sooglossidae from the Seychelles), implying relict survival in the two respective areas (Bossyut and Milinkovitch, 2001).

Roelants et al. (2004) have shown on the basis of molecular phylogenies that the Indian subcontinental plate may have had the necessary permissive ecology of humid tropical rainforests and the isolation as an island subcontinent for the origination and preservation of ranid lineages.

6.3.3 Reptiles

6.3.3.1 Lizards

The lacertilian assemblage of the DTASS has not received the attention it deserves mainly because the material is too fragmentary for a realistic systematic analysis. In one of the most comprehensive accounts from several known localities in the DTASS, Rana (2005) has reported the presence of the following Families: Iguanidae; Agamidae; Scincidae; Xenosauridae (Anguiodea). The presence of these families on the drifting plate have considerable significance with regard to the evolution of lizards and their palaeobiogeographical implications, and therefore it is important that additional more complete material becomes available so that there is more certainty in taxonomic identification. According to Rana (2005), the presence of Iguanids which are also known from the Bauru Formation (Late Cretaceous) of Brazil may be the result of migrations via Madagascar with which landmass India shares several other common taxa in the Late Cretaceous including titanosaurid and abelisaurid dinosaurs and sudamericid gondwanathere mammals. The origin and evolution of the Agamidae is again open to question mainly because the data base in the Cretaceous is not very good. It is noteworthy that the oldest lizard (Agamidae) comes from the Carnian of India (Datta and Ray, 2006) again indicating Gondwanan affinities. The Scincomorpha and the Anguimorpha are more cosmopolitan in their distribution at this time and therefore their occurrence in the DTASS does not have the same significance as the iguanids and the agamids. The Family Anguidae known by vertebrae from the Intertrappean beds is believed to have originated in Laurasia (Prasad and Rage, 1995).

6.3.3.2 Snakes

Snakes are not uncommon in the DTASS and have now been reported from several localities ranging along the peripheral Deccan Trap flows. The first reported fossil snake was a Boidae from the classic localitiy of Pisdura and was obtained by screen washing (Jain and Sahni, 1983). Subsequently, snakes have been reported from the mammal yielding locality of Naskal and Asifabad, Andhra Pradesh (Prasad and Sahni, 1988; Rage and Prasad, 1992), from Takli near Nagpur (Gayet et al., 1984) from Kelapur and Anjar in Kachchh (Rage et al., 2003). The snake taxa include *Indophis* (?Nigerophiidae), Madtsoiidae, and cf. *Coniophis* (Aniliidae).

According to Rage et al. (2003, 2004), the madtsoiid and *Indophis* have a Gondwanan affinity but the relationships of the other snakes is not that certain mainly because they are based on meagre and poorly preserved material. Rage et al. (2003) have recently documented an interesting snake assemblage from the Lower Eocene Panandhro Lignite Mines of Kachchh (Palaeophiidae, ?Madtsoiidae or Boidae and the Colubroidae). Palaeophiids are the most common with the earliest representation of a near-shore aquatic taxon *Pterosphenus*. At present it is not possible to distinguish whether some other poorly preserved vertebrae belong to the ?Madtsoiidae or the Boidae. However, some material has been assigned to the Family Colubridae (Rage et al., 2004) which would then record the earliest Cenozoic occurrence of this family in Asia (Rage et al., 2004).

6.3.3.3 Turtles

Jain (1986) described some good material of turtles from the Lameta Formation of Pisdura, namely the podocnemidid *Shweboemys pisdurensis*. *Shweboemys* has also been reported from the Neogene of Myanmar and Pakistan. Recently, Gaffney et al. (2001, 2003, 2006) have revised the taxonomic status of two genera of side-necked turtles described from two widely separated Maastrichtian localities in peninsular India. These are the pelomedusid pleurodires *Kurmademys* (tribe Kurmademydini, Family Bothremydidae) from the Kallamedu Formation of southern India near the village of Kallamedu, Tamil Nadu, and *Sankuchemys sethnai* from the Mumbai Intertrappeans (Green Tuff level, Amboli Quarry, Jogeshwari). *Kurmademys kallamedensis,* is represented by a skull. *Kurmademys* is a sister taxon of *Sankuchemys,* the two sharing several unique features. The tribe Kurmademydini is close to the Tribe Cerachelyini which is known from the Santana Formation of Brazil and the Cenomanian Kem Kem sediments of Morocco. Another closely related tribe Bothremydini has several taxa distributed in the Campanian-Maastrichtian of Europe, Egypt, Morocco and the eastern United States.

6.3.3.4 Crocodiles

It is quite surprising considering the good preservation and better representation of crocodiles from other parts of the globe that Maastrichtian crocodiles from the DTASS are known mainly from isolated teeth or vertebrae and therefore that their taxonomy has been a problem. The presence of dyrosaurid crocodiles is based mainly on isolated vertebrae first reported from the subcontinental plate in the Eocene by Buffetaut (1978). Rana and Sati (2000) also referred to "alligatorine" affinities of some other material they had collected from the DTASS. This material is now believed to be close to *Hamadasuchus* which was described by Buffetaut (1994) from the mid-Cretaceous sediments of Morocco and *Mahajungasuchus* from the Maevarno Formation (Maastrichtian) of Madagascar (Prasad and De Lapparent De Broin, 2002).

In their most recent assessment of the crocodiles, Prasad and De Lapparent De Broin (2002), have left the taxonomic affinity and relationships rather open in the absence of cranial material.

6.3.3.5 Dinosaurs

The Cretaceous dinosaurs of India are now better known than they were about two decades ago but in terms of global standards they are still too poorly known to be commented upon (Jain and Bandyopadhyay, 1997; Sahni, 2006; Wilson et al., 2003). The best known Lameta dinosaurs are the titanosaurid and abelisaurids. Several other forms are known (Chatterjee and Scotese, 1999) but these still await detailed description. What is clear is that the affinities of the Indian Maastrichtian dinosaurs are with those of South America and Madagascar, and that the titanosaurids and their eggshell morpho-structural types are similar to those found in and around the Iberian peninsular. Wilson et al. (2003) described a new abelisaurid dinosaur *Rajasaurus* and discussed its affinities with *Majungatholus* from Madagascar.

6.3.4 Mammals

Mammals recovered from the DTASS are becoming slowly better known but their affinities are still problematical mainly because most of the material is represented by assorted isolated teeth while even jaws are relatively few (Das Sarma et al., 1995; Gheerbrant, 1990; Khajuria and Prasad, 1998; Khosla et al., 2004; Krause et al., 1997; Prasad, 2005; Prasad and Sahni, 1988; Prasad et al., 2007; Rana and Wilson, 2003). However, it is becoming apparent that with time and better prospecting the record will get much better and that it will lead to greater insights into the origins, dispersals and evolution of placental mammals and the gondwanatheres (Prasad, 2005).

At present the mammalian assemblage is dominated by five main taxa, *Deccanolestes, Sahnitherium*; *Kharmerungulatum* (an archaic ungulate), ?Otolestidae; a gondwanathere, *Bharattherium*, with close affinity to the sudamericid from Madagascar. Though *Deccanolestes* has often been compared to *Abotylestes* from the Late Palaeocene of Morocco (McKenna, 1995), the similarities are in plesiomorphic characters only (Prasad, 2005). Latest phylogenetic analysis of early mammals has placed the Late Cretaceous placental mammals (*Deccanolestes*) of India in the Laurasian clade (Wible et al., 2007). The gondwanathere *Bharattherium bonapartei* (Prasad et al., 2007) on the other hand, is certainly more interesting for palaeobiogeographic reasons. So far, gondwanatheres are restricted to the southern continents and are known from the Late Cretaceous-Palaeocene sequences of South America, and the Late Cretaceous sediments of Madagascar. The Madagascan taxon *Lavanify* is considered to be the "sister taxon" of *Bharattherium*. It is of particular interest that the group appears to have been fairly widespread in the "southern continents" as some new material recovered from Tanzania by Krause et al. (2003) appears to have gondwanatheran affinity, and there is also a recent report of a gondwanathere from the Eocene of Antarctica (Requero et al., 2002). The presence of an archaic Late Cretaceous ungulate on the Indian plate also raises several issues related to palaeobiogeography.

6.3.5 Associated Fossil Floras and Invertebrates

Palynological assemblages dominated by *Aquilapollenites, Gabonispori* and *Azolla* are common throughout the DTASS (Baksi and Deb, 1976; Fredrikson, 1994; Prakash et al., 1990; Sahni et al., 1996); such occurrences being recorded initially from the Bengal Basin subcrops as well as in deep oil exploration wells off the east coast (Baksi and Deb, 1976; Sahni et al., 1996). In one exceptional case, a typical Palaeocene pollen assemblage has been recorded from the Lalitpur Intertrappeans in Uttar Pradesh with forms that are also known from the Madh Formation of Kachchh. The megafloral diversity in the DTASS has been studied for over six decades and is truly extraordinary. However, the plant lineages have not been studied from an evolutionary viewpoint focussed on centres of origin and the geodynamic earth framework and therefore their palaeobiogeographic significance is not fully known.

Ambwani et al. (2003) reported the occurrence of freshwater diatoms from outcrops as well as from dinosaurian coprolite contents. This constitutes the oldest global record of freshwater diatoms. Also from the DTASS at Daiwal, Maharashtra; Samant and Mohabey (2003) recorded fresh-water dinoflagellate cysts associated with a non-marine fish fauna that included the gar fish, *Lepisosteus* in lacustrine environments (Mohabey, 1996).

Monocots, in particular palm lineages are very common from the DTASS (Mehrotra, 2003). Recently, the presence of grasses represented by five taxa of living grass subclades (Poaceae) were reported from dinosaurian coprolite contents (Prasad et al., 2005). This is the oldest record of grasses globally and shows that even at the end Cretaceous, grasses were quite diversified. The occurrence of grasses in the Maastrichtian of India can also be seen in the light of early adaptation of gondwanathere mammals for grazing by hypsodonty.

Recently, several workers have focussed their attention on the ostracode assemblages from the DTASS (Whatley and Bajpai, 2000a) and their revision has resulted in a re-thinking of the relationship of the ostracodes at the species level. It is now believed that the majority of freshwater ostracode species are highly endemic although at the generic level affinities do exist with a few forms known elsewhere including Asia. Therefore the ostracodes as presently known would point to diversification in isolated environments.

6.4 Discussion

Tracing the drift history of India from its separation from the Gondwanaland continents to its collision with Asia is a challenging task (Briggs, 2003; Krause and Maas, 1990; Sahni and Kumar, 1974; Sahni et al., 1982). The exercise seeks to integrate spatial data sets and temporal time scales with the geophysical and the biological data, in order to account for the evolution, origination, dispersal and extinction of faunas and floras during the major mass extinction at the Cretaceous-Tertiary Boundary. This event was also coincident with one of the most extensive continental flood basalt eruptions known. During the later period of the drift, the Palaeocene-Eocene Thermal Maxima (PETM) event coincided with and may have acted as a trigger for the origination of several new forms, some of which have been recorded from the Indian raft (Sahni et al., 2006 and references therein). Although the sea floor spreading data is well mapped out, the fact that the northern margin of Greater India cannot be accurately demarcated makes the issues related to biotic affinities all that more difficult. For example, at present it is not possible to validate or invalidate with certainty the hypothesis that Greater India may have served as a conduit for Gondwanan elements to board intervening microcontinents and island arc systems in the Neotethys Ocean (McKenna, 1973) *before* the northern promontory of the Indian subcontinental plate reached the southern margin of Asia at the KT boundary or slightly later (Jaeger et al., 1989). At best one can analyze the current data set for biotic relationships throughout the history of the

Indian raft and speculate on the affinities of the biotas during India's northward flight.

It was Krause and Maas (1990) who initially surmised that India may have been ideally suited for the origin of new lineages and their subsequent dispersal. Recently, based on the estimated ages of the lineages of modern frog taxa from the Seychelles microcontinent and the Western Ghat region of India, Bossyut and Milinkovitch (2001), proposed the "Out of India" hypothesis which has since been extended to include various other faunal and floral lineages (Karanth, 2006). In the following section, a brief summary is presented about the question of endemism and origination expected on an island subcontinent lacking stable corridors of dispersal for large terrestrial elements.

In an island context, one usually expects a high degree of taxic endemism and origination. The DTASS record the oldest occurrence of grasses, diversified into five taxa of living subclades belonging to the Poaceae and found in the coprolitic material of titanosaurid dinosaurs at the classic locality of Pisdura (Prasad et al., 2005). In addition, the oldest record of fresh water diatoms is also found in the Lameta sediments as well as in the dinosaurian coprolites (Ambwani et al., 2003). At the specific level, Whatley, Bajpai and co-workers have demonstrated in several papers from widely distributed DTASS that the ostracodes are endemic except for a few forms (Whatley and Bajpai, 2000b).

During the same time period, data for leptodactylid frogs, gondwanathere mammals and abelisaurid and titanosaurid dinosaurs suggest an unequivocal affinity to South America and Madagascar. Do these Maastrichtian elements reflect survival as relicts from a Gondwana – inherited biota when India was attached to Gondwanaland or do they represent continued physical contact with the southern continents at the Cretaceous-Tertiary Boundary? The present data cannot reliably resolve this issue.

As the Palaeocene has not yielded very many diagnostic faunal elements, knowledge of the early Tertiary faunas relies on the better known Lower Eocene vertebrates recovered from the Early Tertiary Lignite/Coal Measures (Bajpai et al., 2005; Rana et al., 2004, 2005, 2006; Rose et al., 2006; Sahni et al., 2004, 2006; Samant and Bajpai, 2001; Thewissen et al., 1996, 2001). There are several cases of endemic forms reported from the Lower Eocene including the now classic example of the earliest whales evolved from a terrestrial form (Bajpai and Gingerich, 1998; Thewissen et al., 1996, 2001). Three other mammalian groups are the endemic anthracobunids, considered to be the sister group of proboscideans (Rose et al., 2006), the sirenians (Bajpai et al., 2006) and the chiropterans represented by one of the most diverse Lower Eocene bat faunas known anywhere in the world (Smith et al., 2006). First occurrences are not confined to terrestrial elements but also include marine fish, such as gobies (Bajpai and Kapur, 2004).

After the collision with Asia, the Indian fauna exhibits the presence of several forms that had been previously recorded in Asia (Kumar, 2001; Kumar and Loyal, 1987; Kumar and Sahni, 1985; Sahni and Kumar, 1974). According to the latter author, these migrants include brontotheres, tapiroids, and hyracodonts (Perissodactyla); rodents; dichobunids (Artiodactyla); hyaenodonts (Creodonta) and

Primates. In addition, he states that there are some forms reported from Pakistan but which still have not been reported from India, namely, ?Esthonychidae (Tillodontia), Arctocyonidae (Arctocyonia), and Plesiadapiformes (Primates). Thewissen et al. (2001) has provided a fairly complete faunal list of such forms.

Acknowledgments This article is financially supported by an INSA senior scientist project no: INSA-SP/SS/2006/2841 to the author who would also like to thank Drs. Caroline Stromberg and Vandana Prasad for their help in drafting Fig. 6.1. He further acknowledges the detailed efforts of the reviewers in making general improvements of the article.

References

Ambwani, K, Sahni, A, Kar, R, Dutta, D (2003) Oldest known non-marine diatoms (*Aulacoseira*) from the Deccan Intertrappean Beds and Lameta Formation (Upper Cretaceous of India). Rev Micropaleontol, 46:67–71.

Bajpai, S, Gingerich, PD (1998) A new Eocene Archaeocete (Mammalia, Cetacea) from India and the time of origin of whales. Proc Natl Acad Sci, 95:5464–5468.

Bajpai, S, Kapur, VV (2004) Oldest known gobiids from Vastan Lignite Mine (Early Eocene), Surat. Curr Sci, 87:433–435.

Bajpai, S, Kapur, VV, Das, DP, Tiwari, BN, Saravanan, N, Sharma, R (2005) Early Eocene land mammals from the Vastan Lignite Mine, District Surat (Gujarat), western India. J Palaeontol Soc India, 50:101–113.

Bajpai, S, Thewissen, JGM (2002) Vertebrate fauna from Panandhro Lignite field (Lower Eocene), district Kachchh, western India. Curr Sci, 82:507–508.

Bajpai, S, Thewissen, JGM, Kapur, VV, Tiwari, BN, Sahni, A (2006) Eocene and Oligocene Sirenians (Mammalia) from Kachchh, India. J Vertebr Paleontol, 26:400–410.

Baksi, SK, Deb, U (1976) On new occurrence of *Aquilapollenites* from Eastern India. Pollen Spores, 18(3):399–406.

Bardhan, S, Gangopadhyay, TK, Mandal, U (2002) How far did India drift during the Late Cretaceous?-*Placenticeras kaffrarium* Etheridge 1904 (Ammonoidea) used a measuring tape. Sed Geol, 147:193–217.

Biju, SD, Bossuyt, F (2003) New frog family from India reveals an ancient biogeographical link with the Seychelles. Nature, 425:711–714.

Bossyut, F, Milinkovitch, MC (2001) Amphibians as indicators of early Tertiary 'out of India' dispersal of vertebrates. Science, 291:93–95.

Briggs, JC (2003) The biogeographic and tectonic history of India. J Biogeogr, 30:381–388.

Buffetaut, E (1978) Crocodilians from the Eocene of Pakistan. N J Geol Palaeontol Abh, 156:262–283.

Buffetaut, E (1994) A new crocodilian from the Cretaceous of Southern Morocco. C R Acad Sci, 19:1563–1568.

Cappetta, H (1972) Les poisson Cretaces et Tertiaires Du Basin des Iullemmeden (Republic du Niger). Palaeovertebr Montp, 5:179–251.

Chatterjee, S, Scotese, LR (1999) The breakup of Gondwana and the evolution of the Indian plate. Proc Indian Natl Sci Acad, 65A:397–425.

Cione, AL, Prasad, GVR (2002) The oldest known catfish (Teleostei: Siluriformes) from Asia (India, Late Cretaceous). J Paleontol, 76:190–193.

Clyde, WC, Khan, IH, Gingerich, PD (2003) Stratigraphic response and mammalian dispersal during initial India-Asia collision: evidence from the Ghazij Formation. Baluchistan, Pakistan. Geology, 31:1097–1100.

Das Sarma, DC, Anantharaman, S, Vijayasarathi, G, Nath, TT, Rao, CV (1995) Palaeontological studies for the search of micromammals in the infra- and inter-trappean horizons of Andhra Pradesh. Rec Geol Surv India, 128:223.

Datta, PM, Ray, S (2006) Earliest lizard from the Late Triassic (Carnian) of India. J Vertebr Paleontol, 26:795–800.

Fredrikson, NO (1994) Middle and Late Palaeocene angiosperms pollen from Pakistan. Palynology, 18:91–137.

Gaffney, ES, Chatterjee, S, Rudra, DK (2001) *Kurmademys,* a new side-necked turtle (Pelomedusoides: Bothremydidae) from the Late Cretaceous of India. Am Mus Novit, 3321: 1–16.

Gaffney, ES, Sahni, A, Schleich, H, Singh, SD, Srivastava, R (2003) *Sankuchemys,* a new side-necked turtle (Pelomedusoides:Bothremydidae) from the Late Cretaceous of India. Am Mus Novit, 3405:1–10.

Gaffney, ES, Tong, H, Meylan, PA (2006) Evolution of the side-necked turtles: The Families Bothremydidae, Euraxemididae and Araripemydidae. Bull Am Mus Nat Hist, 300:1–698.

Garzanti, E, Critelli, S, Ingersoll, RV (1986) Paleogeographic and paleotectonic evolution of the Himalayan Range as reflected by detrital modes of Tertiary sandstones and modern sands (Indus transect, India and Pakistan). Geol Soc Am Bull, 108:631–642.

Gayet, M, Rage, JC, Rana, RS (1984) Nouvelles ichthyofaune et herpetofaune de Gitti Khadan le plus ancient gisement connu du Deccan (Cretace/Paleocene) a Microvertebres. Implications palaeogeographiques. Mem Geol Soc Fr NS, 147:55–67.

Gheerbrant, E (1990) On the early biogeographical history of African placentals. Hist Biol, 4: 107–116.

Gingerich, PD, Abbas, SG, Arif, M (1997) Early Eocene *Quettacyon parachai* (Condylarthra) from the Ghazij Formation of Baluchistan (Pakistan): oldest Cenozoic land mammal from South Asia. J Vertebr Paleontol, 17:629–637.

Hora, SL (1939) On some fossil fish scales from the Intertrappean Beds at Deothan and Kheri, Central Provinces. Rec Geol Surv India, 73:267–297.

Jaeger, JJ, Courtillot, V, Tapponier, P (1989) Palaeontological view of the Deccan Traps, the Cretaceous/Tertiary boundary and the India-Asia Collision. Geology, 17:316–319.

Jain, SL (1986) New pelomedusid turtle (Pleurodire: Chelonia) from the Lameta Formation (Maastrichtian) of Dongargaon, central India and a review of Pelomedusids from India. J Palaeontol Soc India, 31:63–75.

Jain, SL, Bandyopadhyay, S (1997) New titanosaurid (Dinosauria: Sauropoda) from the Late Cretaceous of central India. J Vertebr Palaeontol, 17:114–136.

Jain, SL, Sahni, A (1983) Some upper Cretaceous vertebrates from central India and their palaeo-geographic implications. Indian Association of Palynostratigraphers, Symposium Birbal Sahni Institute of Palaeobotany, Lucknow, Uttar Pradesh, pp. 66–83.

Karanth, KP (2006) Out of India Gondwanan origin of some tropical Asian biota. Curr Sci, 90: 789–792.

Khajuria, CK, Prasad, GVR (1998) Taphonomy of a Late Cretaceous mammal-bearing microverte-brate assemblage from the Deccan inter-trappean beds of Naskal, peninsular India. Palaeogeogr Palaeoclimatol Palaeoecol, 137:153–172.

Khosla, A, Kapur, VV, Sereno, PC, Wilson, JA, Dutheil, D, Sahni, A, Singh, MP, Kumar, S, Rana, RS (2003) First dinosaur remains from the Cenomanian- Turonian of the Nimar Sandstone (Bagh Beds), District Dhar, Madhya Pradesh, India. J Palaeontol Soc India, 48:115–127.

Khosla, A, Prasad, GVR, Verma, O, Jain, AK, Sahni, A (2004) Discovery of a micromammal yielding Deccan intertrappean site near Kisalpuri, Dindori District, Madhya Pradesh. Curr Sci, 87:380–383.

Khosla, A, Sahni, A (2003) Biodiversity during the Deccan volcanic eruptive episode. J Asian Earth Sci, 21:895–908.

Krause, DW, Gottfried, MD, O'Connor, PM, Roberts, EM (2003) A Cretaceous mammal from Tanzania. Acta Palaeontol Polonica, 48:321–330.

Krause, DW, Maas, MC (1990) The biogeographic origins of Late Paleocene – Early Eocene mammalian immigrants to the western Interior of North America. Geol Soc Am Spec Pap, 243:71–105.

Krause, DW, Prasad, GVR, Koenigswald, WV, Sahni, A, Grine, FG (1997) Cosmopolitanism among Gondwana Late Cretaceous mammals. Nature, 390:504–507.

Kumar, K (2001) Distribution and migration of Paleogene terrestrial mammal faunas in the Indian subcontinent. International Conference on Distribution and Migration Tertiary Mammals in Eurasia, The University of Utrecht, Utrecht, The Netherlands, pp. 29–31.

Kumar, K, Loyal, RS (1987) Eocene icthyofauna from the Subathu Formation, Northwestern Himalaya, India. J Palaeontol Soc India, 32:60–84.

Kumar, K, Sahni, A (1985) Eocene mammals from the upper Subathu Group, Kashmir Himalaya, India. J Vertebr Paleontol, 5:153–168.

McKenna, MC (1973) Sweepstakes, filters, corridors, Noah's Arc and beached Viking funeral Ships in paleogeography. In: Tarling, DH, Runcorn, SK (eds) Implications of continental drift to the Earth Sciences, Vol. 3. Academic Press, New York, NY.

Mckenna, MC (1995) The mobile Indian raft: a reply to Rage and Jaeger. Syst Biol, 44:265–271.

Mehrotra, RC (2003) Status of plant megafossils during the Early Palaeogene of India. Geol Soc Am Spec Pap, 369:413–423.

Mohabey, DM (1996) Depositional environments of Lameta Formation (Late Cretaceous) of Nand-Dongargaon Inland Basin, Maharashtra: the fossil and lithological evidences. In: Sahni, A (ed) Cretaceous stratigraphy and palaeoenvironments – Rama Rao Volume, Mem Geol Soc India, 37:363–386.

Prakash, T, Singh, RY, Sahni, A (1990) Palynofloral assemblage from the Padwar Deccan inter-trappeans (Jabalpur), M. P. In: Sahni, A, Jolly, A (eds) Cretaceous event stratigraphy and the correlation of the Indian nonmarine strata (IGCP 216 and 245). Panjab University, Chandigarh, India, pp. 68–69.

Prasad, GVR (2005) Mammalian perspective of Late Cretaceous palaeobiogeography of the Indian subcontinent. Gondwana Geol Mag Nagpur, 8:111–122.

Prasad, GVR, Cappetta, H (1993) Late Cretaceous selachians from India and the age of the Deccan traps. Palaeontol, 36:231–248.

Prasad, GVR, De Lapparent de Broin, F (2002) Late Cretaceous crocodile remains from Naskal (India): comparisons and biogeographic affinities. Ann Paléontol, 88:19–71.

Prasad, GVR, Khajuria, CK (1995) Implications of the infra- and inter-trappean biota from the Deccan, India for the role of volcanism in Cretaceous-Tertiary boundary extinctions. J Geol Soc Lond, 152:289–296.

Prasad, GVR, Khajuria, CK (1996) Palaeoenvironment of the Late Cretaceous mammal-bearing Intertrappean beds of Naskal, Andhra Pradesh, India. In: Sahni, A (ed) Cretaceous stratigraphy and palaeoenvironments – Rama Rao Volume, Mem Geol Soc India, 37:337–362.

Prasad, GVR, Khajuria, CK, Manhas, BK (1995) Palaeobiogeographic significance of the Deccan infra- and intertrappean biota from peninsular India. Hist Biol, 9:319–334.

Prasad, GVR, Rage, JC (1991) A discoglossid frog in the Late Cretaceous (Maastrichtian) of India: Further evidence for a terrestrial route between India and Laurasia in the latest Cretaceous. C R Acad Sci, 313:273–278.

Prasad, GVR, Rage, JC (1995) Amphibians and squamates from the Maastrichtian of Naskal, India. Cret Res, 16:95–107.

Prasad, GVR, Rage, J-C (2004) Fossil frogs (Amphibia: Anura) from the Upper Cretaceous Intertrappean beds of Naskal, Andhra Pradesh. Rev Paléobiol, 23:99–116.

Prasad, GVR, Sahni, A (1988) First Cretaceous mammal from India. Nature, 332:638–640.

Prasad, V, Strömberg, CAE, Alimohammadian, H, Sahni, A (2005) Dinosaur coprolites and the early evolution of grasses and grazers. Science, 310:1177–1180.

Prasad, GVR, Verma, O, Sahni, A, Krause, DW, Khosla, A, Parmar, V (2007) A new Late Cretaceous Gondwanatherian Mammal from Central India. Proc Indian Natl Sci Acad, 73:17–24.

Rage, J-C, Bajpai, S, Thewissen, JGM, Tiwari, BN (2003) Early Eocene snakes from Kutch, Western India, with a review of the Palaeophiidae. Geodiversitas, 25:695–716.

Rage, JC, Prasad, GVR (1992) New snakes from the Late Cretaceous (Maastrichtian) of Naskal, India. N J Geol Paläontol Abh, 187:83–97.

Rage, J-C, Prasad, GVR, Bajpai, S (2004) Additional snakes from the uppermost Cretaceous (Maastrichtian) of India. Cret Res, 25:425–434.

Rana, RS (1988) Freshwater fish otoliths from the Deccan Traps associated sedimentary (Cretaceous-Tertiary transition) beds of Rangapur, Hyderabad District, Andhra Pradesh, India. Geobios, 21:465–493.

Rana, RS (2005) Lizard fauna from the Intertrappean (Late Cretaceous-Early Palaeocene) beds of Peninsular India. Gondwana Geol Mag Nagpur, 8:123–132.

Rana, RS, Kumar, K, Loyal, RS, Sahni, A, Rose, KD, Mussell, J, Singh, H, Kulshreshtha, SK (2006) Selachians from the early Eocene Kapurdi Formation (Fullers Earth), Barmer District, Rajasthan, India. J Geol Soc India, 67:509–522.

Rana, RS, Kumar, K, Singh, H (2004) Vertebrate fauna from the subsurface Cambay Shale (Lower Eocene), Vastan Lignite Mine, Gujarat, India. Curr Sci, 87:425–427.

Rana, RS, Sati, KK (2000) Late Cretaceous-Palaeocene crocodilians from the Deccan Trap associated sedimentary sequences of peninsular India. J Palaeontol Soc India, 45:123–136.

Rana, RS, Singh, H, Sahni, A, Rose, KD, Saraswati, PK (2005) Early Eocene chiropterans from the Vastan lignites, Gujarat, western peninsular: oldest record of bats from Asia. J Palaeontol Soc India, 50:93–100.

Rana, RS, Wilson, GP (2003) New Late Cretaceous mammals from the intertrappean beds of Rangapur, India and palaeobiogeographic framework. Acta Palaeontol Polonica, 48:331–348.

Requero, MA, Sergio, AM, Santillana, SN (2002) Antarctica Peninsula and South America (Patagonia) Paleogene Terrestrial Faunas and Environments: biogeographic relationships. Palaegeogr Palaeoclimatol Palaeoecol, 179:189–210.

Roelants, K, Jiang, J, Bossuyt, F (2004) Endemic ranid (Amphibia: Anura) genera in southern mountain ranges of the Indian subcontinent represent ancient frog lineages: evidence from molecular data. Mol Phylogenet Evol, 31:730–740.

Rose, KD, Smith, T, Rana, RS, Sahni, A, Singh, H, Missiaen, P, Folie, A (2006) Early Eocene (Ypresian) continental vertebrate assemblage from India, with description of a new anthracobunid (Mammalia, Tethytheria). J Vertebr Paleontol, 26:19–25.

Russell, DE, Gingerich, PD (1981) Lipotyphla, Proteutheria (?), and Chiroptera (mammalia) from the early – middle Eocene Kuldana formation of Kohat (Pakistan). Contr Mus Paleontol Univ Michigan, 25:277–287.

Sahni, A (1999) India-Asia collision: ecosystem changes. J. B. Auden Lecture. Wadia Institute of Himalayan Geology, Dehra Dun, Uttarakhand, pp. 1–27.

Sahni, A (2006) Biotic response to the India-Asia Collision: changing palaeoenvironments and vertebrate faunal relationships. Palaeontographica (Stuttgart, Germany), 278A:15–26.

Sahni, A, Jolly, A (1993) Eocene mammals from Kalakot, Kashmir Himalaya: community structure, taphonomy and palaeobiogeographical implications. Kaupia, 3:209–222.

Sahni, A, Kumar, V (1974) Palaeogene palaeobiogeography of the Indian subcontinent. Palaeogeogr Palaeoclimatol Palaeoecol, 13:209–226.

Sahni, A, Kumar, K, Hartenberger, JL, Jaeger, JJ, Rage, JC, Sudre, J, Vianey-Liaud, M (1982) Microvertbrs nouveau des Traps du Deccan (Inde): mise en vidence d'une voie de communication terrestre probable entre la Laurasie et l' Inde a la limite Crtac-Tertiaire. Bull Soc Gol Fr, 24:1093–1099.

Sahni, A, Rana, RS, Loyal, RS, Saraswati, PK, Mathur, SK, Rose, KD, Tripathi, SKM, Garg, R (2004) Western margin palaeocene-lower Eocene lignite: biostratigraphic and palaeoecological constraints. Proceedings of the 2nd associations of petroleum geology, ONGC, Khajuraho, Madhya Pradesh, pp. 1–18.

Sahni, A, Saraswati, PK, Rana, RS, Kumar, K, Singh, H, Alimohammadian, H, Sahni, N, Rose, KD, Singh, L, Smith, T (2006) Temporal constraints and depositional palaeoenvironments of the Vastan Lignite Sequence, Gujarat: analogy for the Cambay Shale Hydrocarbon Source Rock. Indian J Petroleum Geol, 15:1–20.

Sahni, A, Tandon, SK, Jolly, A, Bajpai, S, Sood, A, Srinivasan, S (1994) Upper Cretaceous dinosaur eggs and nesting sites from the Deccan Volcano- Sedimentary province of Peninsular

India. In: Carpenter, K, Hirsch, KF, Horner, JR (eds) Dinosaur eggs and babies. Cambridge University Press, Cambridge, MA.

Sahni, A, Venkatachala, BS, Kar, RK, Rajanikanth, A, Prakash, T, Prasad, GVR, Singh, RY (1996) New palaeontological data from the Intertrappean beds: implications for the latest record of dinosaurs and synchronous initiation of volcanic activity in India. In: Sahni A (ed) Cretaceous stratigraphy and palaeoenvironments – Rama Rao Volume, Mem Geol Soc India, 37:267–203.

Samant, B, Bajpai, S (2001) Fish otoliths from the subsurface Cambay Shale (Lower Eocene), Surat Lignite Field, Gujarat, India. Curr Sci, 81:758–759.

Samant, B, Mohabey, DM (2003) Palynology study of Late Cretaceous (Maastrichtian) sediments of Nand-Dongargaon area: palaeoclimate and palaeoenvironment. Proceedings of the 19th Indian colloquium on micropalaeontology and stratigraphy: sympsoium on recent development in Indian Ocean palaeoceanography and palaeoclimate, Banaras Hindu University, Varanasi, Uttar Pradesh, pp. 63–64.

Serra-Kiel, J, Hottinger, L, Caus, E, Brobne, K, Ferrandez, C, Jauhari, AK, Less, G, Pavlovec, R, Pignatti, J, Samso, JM, Schaub, H, Sirel, E, Struogo, A, Tambareau, Y, Tosquella, J, Zakrevskaya, E (1998) Larger foraminiferal biostratigraphy of the Tethyan Paleocene and Eocene. Bull Geol Soc Fr, 169:281–299.

Smith, JB, Lamanna, MC, Lacovara, KJ, Dodson, P, Smith, JR, Poole, JC, Geigengack, R, Attia, Y (2001) A giant sauropod dinosaur from an Upper Cretaceous Mangrove deposit in Egypt. Science, 292:1704–1706.

Smith, T, Rana, RS, Rose, KD, Sahni, A (2006) Earliest bats from India. J Vertebr Paleontol (Abs), 26:127A.

Soler-Gijon, R, Martinez, NL (1998) Sharks and rays (Chondrichthyes) from the Upper Cretaceous red beds of the south-central Pyrenees (Lleida, Spain): indices of an India-Eurasia connection. Palaeogeogr Palaeoclimatol Palaeoecol, 141:1–12.

Spinar, ZV, Hodrova, M (1985) New knowledge of the genus *Indobatrachus* (Anura) from the Lower Eocene of India. Amphibia-Reptilia, 6:363–376.

Tandon, SK, Sood, A, Andrews, JE, Dennis, PF (1995) Palaeoenvironment of the dinosaur bearing Lameta Beds (Maastrichtian), Narmada Valley, Central India. Palaeogeogr Palaeoclimatol Palaeoecol, 117:153–184.

Thewissen, JGM, Roe, LJ, O'Neil, JR, Hussain, ST, Sahni, A, Bajpai, S (1996) Evolution of Cetacean osmoregulation. Nature, 381:379–380.

Thewissen, JGM, Williams, EM, Hussain, ST (2001) Eocene mammal faunas from northern Indo-Pakistan. J Vertebr Paleontol, 21:347–366.

Whatley, RC, Bajpai, S (2000a) Further nonmarine ostracoda from the Late Cretaceous Intertrappean deposits of the Anjar region, Kachchh, Gujarat, India. Rev Micropaleontol, 43:173–178.

Whatley, RC, Bajpai, S (2000b) Zoogeographical relationships of the Upper Cretaceous nonmarine ostracoda of India. Curr Sci, 79:694–696.

Wible, JR, Rougier, GW, Novacek, MJ, Asher, RJ (2007) Cretaceous eutherians and Laurasian origin for placental mammals near the K/T boundary. Nature, 447:1003–1006.

Wilson, JA, Sereno, PC, Srivastava, S, Bhatt, DK, Khosla, A, Sahni, A (2003) A new abelisaurid (Dinosauria, Theropoda) from the Lameta Formation, (Cretaceous, Maastrichtian) of India. Contr Mus Paleontol Univ Michigan, 31:1–42.

Chapter 7
The Wandering Indian Plate and Its Changing Biogeography During the Late Cretaceous-Early Tertiary Period

Sankar Chatterjee and Christopher Scotese

7.1 Introduction

The biogeography of Indian tetrapods during the Late Cretaceous-Early Tertiary period provides an unparalleled opportunity to examine the complex ways in which the tetrapods responded to the sequence and timing of the rifting, drifting, and collision of Indian plate. Triassic and Jurassic tetrapods of India have widespread Gondwanan relationships (Chatterjee and Scotese, 1999, 2007). With the breakup of Gondwana, India remained isolated as an island continent, but reestablished its biotic links with Africa, Madagascar, South America, and Asia during the Late Cretaceous (Sahni and Bajpai, 1988; Chatterjee and Scotese, 1999; Khosla and Sahni, 2003). During the Palaeocene, India drifted northward as an island continent, when the tetrapods of Gondwanan heritage presumably evolved in intermittent isolation for 20 Ma and radiated into extraordinary diversity. Endemism ended when the Indian plate collided with Asia at the Palaeocene/Eocene boundary. A northeast migration corridor arose at a time when several groups of newly evolved tetrapods might have dispersed into Asia. India also received many groups of tetrapods from the north during its union with Asia (Russell and Zhai, 1987). A great faunal interchange took place during the Eocene. In this paper we integrate the tectonic evolution of the Indian plate with its biogeography during its long northward journey to test the models of vicariance and geodispersal.

7.2 Late Cretaceous Configuration of the Indian Plate

The configuration of Indian plate changed dramatically during the Late Jurassic and Cretaceous as it broke apart sequentially from Gondwana, East Gondwana, Madagascar, and finally from the Seychelles island (Chatterjee and Scotese, 1999).

S. Chatterjee (✉)
Department of Geosciences, Museum of Texas Tech University, Lubbock, TX 79409, USA
e-mail: sankar.chatterjee@ttu.edu

S. Bandyopadhyay (ed.), *New Aspects of Mesozoic Biodiversity*,
Lecture Notes in Earth Sciences 132, DOI 10.1007/978-3-642-10311-7_7,
© Springer-Verlag Berlin Heidelberg 2010

Most plate reconstructions suggest that India was an isolated and island continent during the Cretaceous for more than 100 million years until it docked with Asia in Early Eocene (Barron and Harrison, 1980; Smith, 1988). Such an extended period of continental isolation should have produced a highly endemic Late Cretaceous vertebrate assemblage. To the contrary, Late Cretaceous Indian vertebrates are cosmopolitan and show faunal similarities with those of South America, Madagascar, Africa, and Europe (Chatterjee and Scotese, 1999; Sahni and Bajpai, 1988; Jaegger et al., 1989; Briggs, 1989). Some elements have been interpreted as Gondwanan whereas others show affinities with Laurasian vertebrates.

To resolve this apparent biogeographic contradiction, various land bridges have been proposed that opened up faunal migration route between India and other continents. For example, the maintenance of a Late Cretaceous "Gondwanan" biota in India implied a prolonged connection with other Gondwana continents via Antarctica-Kerguelen Plateau (Krause et al., 1997), via Arabia-Kohistan-Dras volcanic Arc (this paper), via "Greater Somalia" (Chatterjee and Scotese, 1999) or via Africa-Madagascar (Briggs, 1989). On other hand, to account for the presence of Late Cretaceous Eurasiatic elements among Indian Cretaceous vertebrates argues for a connection to Asia via an early collision (Jaegger et al., 1989), or indirectly via a volcanic island arc system, possibly the Dras volcanics of Ladakh region (Sahni et al., 1987; Sahni and Bajpai, 1988) that may have provided the corridor for the dispersal of Laurasiatic elements.

Chatterjee and Scotese (1999) proposed the "Greater Somalia" model as a continental corridor consisting of Iranian and Afghan blocks between eastern Arabia and northwestern India during the Late Cretaceous (Fig. 7.1). However recent plate tectonic reconstructions (Sharland et al., 2004) suggest that "Greater Somalia" broke up from Africa into Lut and Afghan blocks and arrived at the southern margin of Eurasia during the Late Triassic-Early Jurassic time with the spreading of the Neotethys. Because of this conflict in timing, we reject the existence of "Greater Somalia" during the Late Cretaceous. Here we propose an alternative hypothesis, a volcanic island arc, named Oman-Kohistan-Dras Island Arc that linked India with Africa during the Late Cretaceous.

7.2.1 Oman-Kohistan-Dras Island Arc: The Northern Biotic Link Between Africa and India During Late Cretaceous

Sahni et al. (1987) and Jaegger et al. (1989) suggested an early India-Asia collision at the Cretaceous-Tertiary boundary time to explain strong faunal relationships of Indian Late Cretaceous vertebrates with those of Asia. These authors proposed that the Kohistan-Dras volcanic arc (K-D Arc) probably acted as a dispersal route between India and Asia during the Late Cretaceous time. Although we do not support the early collision scenario, we believe that the Kohistan-Dras volcanic arc might be a viable dispersal route between Arabia and India during the Late Cretaceous time. Here we offer the tectonic evolution of the K-D Arc.

During the Late Cretaceous the Oman island arc collided with the northern margin of Arabia (Fig. 7.2). The obducted and uplifted ocean floor associated

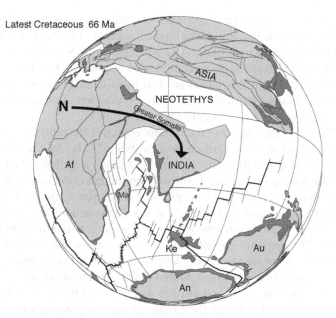

Fig.7.1 Palaeogeographic reconstruction of drifted Gondwana continents during the latest Cretaceous (66 Ma) showing the position of "Greater Somalia" as a continental prong for geodispersal route of Maastrichtian tetrapods (after Chatterjee and Scotese, 1999). In this paper we reject the existence of "Greater Somalia" during the Late Cretaceous but prefer the Kohistan-Dras volcanic Arc as the possible geodispersal route between Africa and India

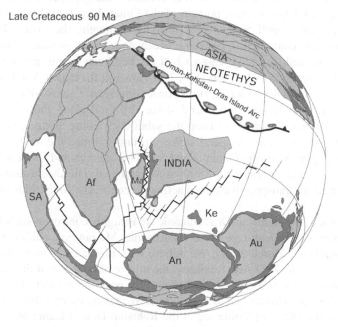

Fig.7.2 Palaeogeographic reconstruction of drifted Gondwana continents during the Late Cretaceous (90 Ma) showing the evolution of Oman-Kohistan-Dras Arc

with this island arc forms one of the most extensive and one of the best exposed ophiolite sequences in the world, the Semail ophiolite of Oman (Glennie et al., 1973; Robertson and Searle, 1990). The outcrops of the Semail ophiolite extend from the Straits of Hormuz to the coast of Arabia near Ra's al Hadd. It is reasonable to assume that originally, this island arc was more laterally extensive and part of the island arc continued eastward from Arabia in the form of Oman-Kohistan-Dras island Arc (Clift et al., 2000; Stampfli and Borel, 2002).

Figure 7.2 shows the closure of the western branch of Neotethys and the collision of the Oman Island arc with northern Arabia about 90 Ma. As a result of decades of oil exploration, the stratigraphic section in Oman is well dated (Sharland et al., 2004), and the timing of the initial collision, uplift, and obduction of the ophiolite is well constrained. A thick carbonate platform developed across Arabia during the Jurassic, Early Cretaceous, and through to the Cenomanian. During the Turonian (90 Ma), a widespread unconformity occurs throughout northern Arabia (Sharland et al., 2004). This regional unconformity may have been due to uplift associated with the southward migration of the peripheral bulge produced during the initial phase of collision between northern Arabia and the Oman island arc.

In northeast Arabia, uplift and erosion was followed by the stacking of thrust sheets (Hawasina Group) during the Coniacian and early Santonian (85 Ma), and the obduction of the Semail ophiolite during the late Santonian-early Campanian (80 Ma). Uplift and unroofing in the late Campanian (75 Ma) was followed by subsidence and the re-establishment of a broad carbonate platform in the Maastrichtian (70 Ma) (Warburton et al., 1990).

To the east of Arabia, the following scenario, though speculative, may have taken place. After the western portion of the Oman Island arc had collided with Arabia and draped itself along Arabia's northern margin, the eastern half of the Oman island arc continued to consume ocean floor in the western Indian Ocean. Due to trench "roll-back" the eastern half of the Oman island arc moved southward along eastern coast of Arabia (Fig. 7.2). However the trench "roll back" would be less likely especially in a near orthogonal convergence across the trench. More likely scenario would be a left-lateral strike slip fault along the eastern boundary of the Arabian Peninsula to keep the subduction going along the trench. The Masirah fracture zone, a left-lateral strike-slip fault, formed the western boundary of remaining portion of the Oman island arc. The Masirah and the Ras Madrakah ophiolites (Shackleton and Ries, 1990) were obducted onto the eastern margin of Arabia as a result transpressive forces acting perpendicular to the Masirah fracture zone.

The origin of the Kohistan-Dras volcanic arc is highly controversial: whether this arc of continental Andean-type or an intraoceanic, Marian-type edifice (Clift et al., 2000; Stampfli and Borel, 2002). We believe both mechanisms might have played in the genesis of the K-D Arc. Similarly, the timing of collision of this arc with the Eurasian margin (Karakoram block) has equally been the subject of great debate, ranging from Late Cretaceous to Palaeocene. We speculate that the eastward continuation of the Oman ophiolite arc is the Kohistan-Dras volcanic arc, which was generated in an intraoceanic setting during Early Cretaceous time, and accreted to the leading edge of the northward drifting Indian plate during the Late Cretaceous time. Subsequently, the arc became accreted to Eurasia during Palaeocene time with the closure of the Neotethys. The K-D Arc is a relict of a thick pile of Late Jurassic to Late Cretaceous forearc basin consisting of basalts, dacites, volcanoclastic sediments, pillow lavas, minor radiolarian cherts, and fossiliferous limestone inclusions (Sengor, 1990). Apparently, these rocks originated in deep water, adjacent to an active, suduction-influenced volcanic terrain during the Early Cretaceous (Clift et al., 2000). Today, the K-D Arc forms a part of the Indus Suture Zone in the Kashmir Himalaya and is rimmed by the Shyok Suture in the north and the Indus Suture in the south (Fig. 7.5). The K–D Arc was initiated during the Early Cretaceous as an eastward extension of the Oman Arc and comprises three main structural and stratigraphic units, the Suru, Naktul, and Nindam Formations from west to east, which has been interpreted as forearc basin (Clift et al., 2000). The arc was erected above a northward-dipping subduction zone. This Oman-Kohistan-Dras Arc may have provided the crucial biotic link between Africa and India during the Late Cretaceous.

During the Cenomanian (~95 Ma), India rifted away from Madagascar and headed northward at a very high speed of 18–20 cm/year covering a distance of about 6,000 km, and then slowed to 5 cm/year during early Eocene after continental collision with Asia (Patriat and Achache, 1984). We project that northward moving Indian plate with a Neotethyan ocean floor at its leading edge would have collided and subducted under the Kohistan-Dras trench sometime during the Late Campanian (75 Ma) or Early Maastrichtian (70 Ma) and formed a series of islands which became accreted to Greater India (Fig. 7.3). The collision of India and the Kohistan-Dras island arc during the Late Cretaceous would have closed the biotic circuit between India and Africa, allowing dinosaur and other terrestrial vertebrate faunas to migrate between the two continents. As India began to move northward, the K-D Arc, severed from the Oman Arc along the Oman-Chaman Transform fault, travelled with India. The oceanic lithosphere of Neotethys between India and Asia was consumed and subducted along the Shyok Suture since the Late Cretaceous to form the Andean-type collision. The K–D Arc was accreted to the southern margin of Asia along the Shyok Suture Zone during the Palaeocene (60 Ma) when it switched from intraoceanic to continental-arc volcanism. It is generally believed that the collision of the K-D Arc with Asia along the Shyok Suture occurred prior to India-Eurasia collision along the Indus Suture (Van der Voo, 1993; Clift et al., 2000).

Here we outline eight distinct stages in the tectonic evolution of the Kohistan-Dras Arc during the initial India-Asian collision: (1) the formation of the K-D Arc in Early Cretaceous time as an eastward continuation of the Oman Arc when India was farther south; (2) as India moved rapidly northward during the Late Cretaceous, the intervening oceanic lithosphere was obducted and uplifted in a form of series of islands north of the K-D Arc, which became an integral part of Indian plate around 70 Ma and formed a dispersal route with Africa (Fig. 7.3); (3) subsequent split of the Kohistan-Dras Arc from the Oman Arc in the Late Cretaceous-Early Palaeocene along the Oman-Chaman Transform Fault, the former attached to India,

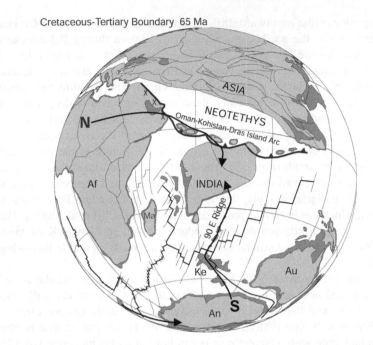

Cretaceous-Tertiary Boundary 65 Ma

Fig.7.3 Palaeogeographic reconstruction of drifted Gondwana continents during the latest Cretaceous (66 Ma) showing two geodispersal routes for migration of Maastrichtian tetrapods between India and other Gondwana continents. The northern route is via Oman-Kohistan-Dras Island Arc to Africa. The southern geodispersal route is via Ninetyeast Ridge-Kergulen Plateau-Antarctica to South America

the latter to Arabia; (4) during the next 20 million years in the Palaeocene time, Indian plate glided rapidly at an accelerated rate of 18–20 cm/year northward between two great transform faults, the Ninetyeast Ridge on the east and Oman-Chaman fault on the west resulting in the subduction of the Neotethyan oceanic lithosphere along the Shyok Suture (Fig. 7.8); (5) collision of the K-D Arc with Eurasia in Late Palaeocene (~60 Ma) along the Shyok Suture; (6) first contact between continental crustal blocks of India and Asia during early Eocene; India-Eurasia collision created a new subduction zone south of the Shyok Suture, the Indus Suture Zone, when continental crust of Greater India began to subduct under Tibet (Fig. 7.4). Continental lithospheres can only be subducted if the slab-pull forces can overwhelm the natural buoyancy of the continental lithosphere. (7) With continued convergence, subduction, and overthrustings of Indian plate, the K-D Arc was emplaced within the Indus Suture Zone (Fig. 7.5); (8) slab-pull forces were sufficiently strong along the Indus Suture to subduct over 1,600 km northern margin of Greater India beneath the Tibetan plateau; continental collision created the rise of the Himalayas and gigantic east–west trending strike-slip faults that are forcing eastern Asia to slide to the east, out of India's path (Fig. 7.5) (Molnar and Tapponnier, 1977).

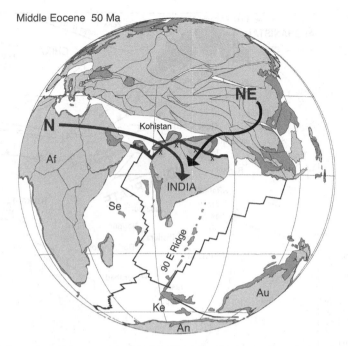

Fig. 7.4 Palaeogeographic reconstruction showing the position of India and other Gondwana continents during the middle Eocene (50 Ma) when India made the initial collision with Asia on its northward journey. A new northern corridor through the Kohistan-Dras Arc allowed "Great Faunal Interchange" between India and Asia

7.2.2 Rajmahal-Kerguelen Hotpot Trail: The Southern Biotic Link Between India and South America During Late Cretaceous

The Ninetyeast Ridge, the longest and most spectacular tectonic feature in the eastern Indian Ocean, becomes progressively younger from north (80 Ma) to the south (40 Ma). It has been proposed that that the Rajmahal Traps of eastern India (117 Ma), the Ninetyeast Ridge, and the Kerguelen active volcano form a hotspot trail as India moved northward during the Cretaceous (Duncan, 1981). Sampson et al. (1998) postulated a subaerial link between India-Madagascar and Antarctica across the Kerguelen Plateau that persisted as late as 80 Ma. This southern dispersal route (Fig. 7.3) might account for the great similarity of the Late Cretaceous terrestrial biota between South America and Indo-Madagascar (via Antarctica).

7.3 Late Cretaceous Biogeography

Terrestrial tetrapod fossils have long been regarded as an important tool in understanding the past distribution of continents (Hallam, 1972). Currently two competing biogeographical principles are considered important to explain the past

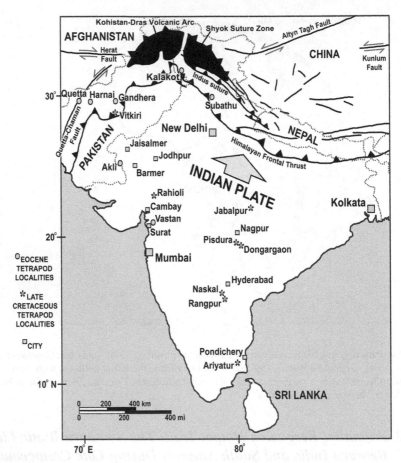

Fig. 7.5 Fossil localities of Late Cretaceous-Early and Middle Eocene deposits in India-Pakistan region showing the major structural features of the Himalayan belts. Note the present position of the Kohistan-Dras volcanic arc, bounded between the Shyok Suture in the north and the Indus Suture in the south

and present distributions of terrestrial tetrapods in the framework of plate tectonics and phylogeny: vicariance and geodispersal (Nelson and Platnik, 1981; Lieberman, 2000, 2003; Upchurch et al., 2002). Vicariance biogeography invokes area separation, driven by continental breakup and emplacement of ocean barrier that fragment and differentiate a once cosmopolitan fauna leading to genetic isolation and morphological divergence. Geodispersal biogeography, on the other hand, is formed by area coalescence, the formation of a connection between two previously geographically isolated areas related to fall of geographic barriers, driven by tectonism or climate change that allows cross migration leading to more homogeneous and uniform tetrapods. The episodes of geodispersal could be formed by large-scale plate tectonic events such as collision of two continents, development of hotspot trails, or formation of island arcs because they eliminate geographic barriers and facilitate concurrent range of expansion of species in many independent clades (Lieberman, 2000, 2003). The role of geodispersal in structuring biodiversity has until recently been severely underestimated and that of vicariance has been overestimated. Recent palaeobiogeographic study indicates that the history of Indian clades and tetrapods involves cycles of both vicariance and geodispersal.

As an index of faunal similarity between two regions, Simpson (1947) suggested a simple formula of $100C/N^1$, where C is the number of faunal units of given taxonomic rank (family level here) common to two areas, and N^1 is the total number of such units in the smaller of the two faunas. In our discussion N^1 represents the total number of taxa present in India.

7.3.1 The Late Cretaceous Tetrapods

Tetrapods are more common in the uppermost Cretaceous continental sediments that are exposed along the fringe of the Deccan Trap province in peninsular India (Fig. 7.5), occurring either beneath the Deccan lava flows (Infratrappean) or as thin intercalations within the basal flows (Intertrappean). These Deccan volcano-sedimentary sequences are highly fossiliferous and contain diverse biotic assemblages (Fig. 7.6).

Diverse tetrapod fossils including frogs (Noble, 1930; Sahni et al., 1982; Prasad and Rage, 1991), turtles (Jain, 1986; Gaffney et al., 2001, 2003), squamates (Chatterjee and Scotese, 2007), crocodiles (Wilson et al., 2001; Rana and Sati, 2000), dinosaurs (Huene and Matley, 1933; Chatterjee and Rudra, 1996; Jain and Bandyopadhyay, 1997; See Chap. 3 by Novas et al., this volume), and mammals (Prasad and Sahni, 1988; Khajuria and Prasad, 1998; Prasad et al., 1994, 2007; Rana and Wilson, 2003) have been recorded from the Infra- and Intertrappean beds and are listed in Table 7.1. Some of the representatives of the Late Cretaceous tetrapod fossils are shown in Fig. 7.7.

7.3.2 Biogeography of the Late Cretaceous Tetrapod Faunas

The greatest incongruence between the palaeobiogeography and India's plate reconstruction during the Late Cretaceous could be resolved once we accept the northern and southern geodispersal routes for faunal migration – one between India and South America via Ninetyeast Ridge-Kerguelen-Antarctica, the other between India and Africa via K-D Arc. Using Fig. 7.3 as a tectonic framework, the biogeography of the Late Cretaceous tetrapod faunas from India can be discussed in relation to other landmasses. The Neotethys was still a barrier for tetrapod migration during the Maastrichtian, but developed three dispersal routes between Laurasia and Gondwana (Rage, 1988; Chatterjee and Scotese, 1999): (1) between North and South America in both directions (ended during the Palaeocene); (2) between

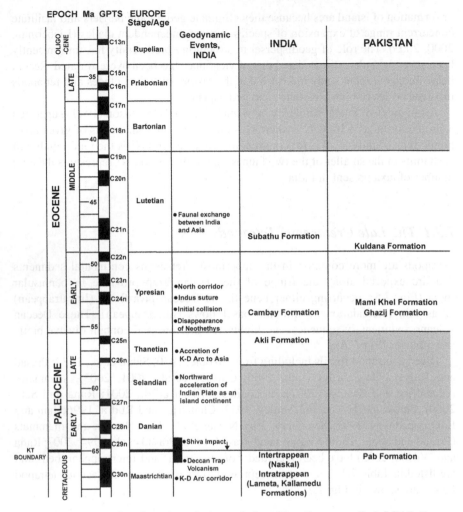

Fig. 7.6 Palaeomagnetic and geochronologic correlation of Late Cretaceous-Early/Middle Eocene continental deposits of India and Pakistan

Eurasia and Africa in both directions (ended by the Eocene), and (3) between India and Africa (ended during the Palaeocene).

The faunal list of Late Cretaceous Indian tetrapods (Table 7.1) includes 16 family level taxa, of which eight of them are present in South America and Laurasia (faunal index 50%), six families are present in Madagascar (faunal index 38%), and four families are present in Africa (faunal index 25%). Interestingly, five families (faunal index 31%) may be endemic or linked to Indian origin. The index of faunal similarity is very high between India-South America and India-Europe during the Maastrichtian indicating the presence of northern and southern biotic links.

Table 7.1 Comparisons of Late Cretaceous Tetrapod Faunas of Indian subcontinent (including Pakistan) with Madagascar, Africa, South America and Laurasia for palaeobiogeographic relationships

	Madagascar	Africa	S. America	Laurasia
AMPHIBIA				
Frogs				
Myobatrachidae			+	
Indobatrachus pusillus				
Pelobatidae				+
Discoglossidae				+
Hylidae				+
REPTILIA				
Turtles				
Bothremydidae	+	+	+	+
Kurmademys kallamedensis				
Sankuchemys sethnai				
Podocnemididae	+	+	+	+
Shweboemys pisdurensis				
Squamates				
Anguidae*				
Nigerophiidae*				
Indophis sahni				
Crocodiles				
Baurusuchidae			+	
Pabwehshi pakistanensis				
Alligatoridae				+
Dinosaurs				
Titanosauria				
Saltasauridae	+		+	+
Isisaurus colberti				
Jainosaurus septentrionalis				
Abelisauridae	+	+	+	
Rajasaurus narmadensis				
Rahiolisaurus gujaratensis				
Noasauridae	+	+	+	
Laevisuchus indicus				
MAMMALIA				
Eutheria				
Archonta				
Family *incertae sedis**				
Deccanolestes hislopi				
D. robustus				
Boreosphenida				
Family *incertae sedis**				
Sahnitherium rangapurensis				
Sudamericidae	+		+	
Condylartha				
Family *incertae sedis**				
Kharmerungulatum vanvaleni				

Faunal Index: *31% endemic or Indian origin; Africa – 25%;
Madagascar – 38%; S. America – 50%; Laurasia – 50%.

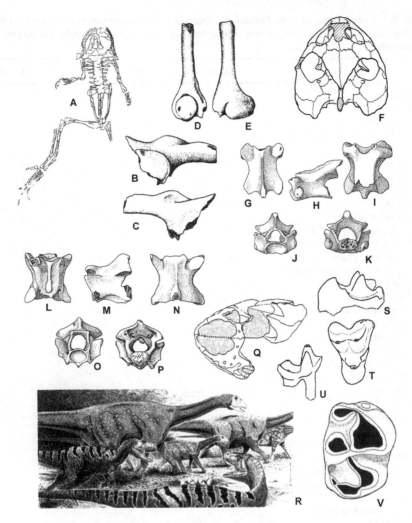

Fig. 7.7 Selected representatives of Late Cretaceous tetrapods from India. (**a–e**) Frogs from the Intertrappean beds; (**a**) *Indobatrachus pusillus*, a myobatrichid frog (after Noble, 1930); (**b** and **c**) lateral and medial views of a right ilium of a discoglossid frog (after Prasad and Rage, 1995); (**d** and **e**) ventral and dorsal views of a right humerus of a possible hylid frog (after Prasad and Rage, 1995). (**f**) *Kurmademys kallamedensis*, a bothremydid turtle from the Kallamedu Formation of Tamil Nadu (after Gaffney et al., 2001). (**g–p**) Squamates from the Intertrappean beds (after Prasad and Rage, 1995); (**g–k**) a trunk vertebra showing dorsal, left lateral, ventral, anterior and posterior views of an anguid lizard; (**l–p**) mid-trunk vertebra of *Indophis sahnii*, a nigeropheid snake. (**q**) *Pabwehshi pakistanensis*, a baurusuchid crocodyliform from Pab Formation of Pakistan (after Wilson et al., 2001). (**r**) Life restoration of Maastrichtian dinosaurs from the Lameta Formation. In the foreground are two abelisaurid theropods such as *Rahiolisaurus gujaratensis*, confronting a herd of titanosaurid sauropods such as *Isisaurus colberti* (after Chatterjee and Rudra, 1996). (**s–u**) Dentition of *Deccanolestes robustus*, a mammalian Archonta from the Intertrappean bed; (**s–t**) right molar (M_2) showing mesial and occlusal views; (**u**) left molar (M_1) in lingual view (after Prasad et al., 1994); (**v**) occlusal view of the right molar tooth of *Kharmerungulatum vanvaleni*, the oldest known ungulate from the Intertrappean bed (after Prasad et al., 2007)

7.4 Early Tertiary Biogeography

After the Cretaceous-Tertiary extinction, India carried its impoverished tetrapods like a "Noah's Ark" as an island continent (McKenna, 1973) during its northward journey in splendid isolation leading to allopatric speciation (Fig. 7.8). There is no fossil record of terrestrial Palaeocene tetrapods from the Indian subcontinent. The recently discovered Akli Formation is very promising for yielding Palaeocene tetrapods.

7.4.1 Eocene Tetrapods

The continental vertebrate-bearing horizons of Early Eocene age are known from several regions of Indo-Pakistan region including the Cambay Formation of the Panhandro and Vastan lignite mines of Gujrat of western India and coal-bearing Ghazij Formation of Baluchistan when India began to converge with Asia (Figs. 7.7 and 7.9). The early Middle Eocene mammals are known from the Subathu Formation of the Kalakot area in the State of Jammu and Kashmir, India (Sahni and

Fig. 7.8 Palaeogeographic reconstruction showing the position of India and other Gondwana continents during the Palaeocene (60 Ma). As India drifted northward toward Asia, the Kohistan-Dras Arc began to move with India and split from the Oman Arc along the Oman-Chaman Transform Fault. In this configuration, India became completely isolated to promote allopatric speciation

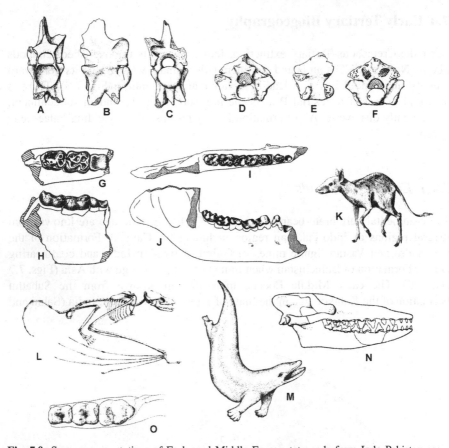

Fig. 7.9 Some representatives of Early and Middle Eocene tetrapods from Indo-Pakistan area. (a–f) Early Eocene snakes from Kutch (after Rage et al., 2003). (a–c) Trunk vertebra of *Pterosphenus kutchensis*, a palaeophid snake showing anterior, left lateral and posterior views. (d–f) Trunk vertebra of a madtsoiid or boid snake showing anterior, left lateral, and posterior views. (g–j) Endemic family of archaic condylarth Quettacyonidae from early Eocene of Balochistan indicating partial isolation of Indian subcontinent (after Gingerich et al., 1999); (g–h) *Quettacyon parachai*, occlusal and right lateral views of right dentary; (i and j) *Machocyon abbasi*, right dentary showing occlusal and lateral views. (k) Life restoration of *Diacodexis*, the oldest known artiodactyls and probably the sister group of whales, known from uppermost Ghazi Formation of Pakistan (after Rose, 2006). (l) The skeleton of early Eocene bat *Icaronycteris sigei* from the Cambay Formation of Gujarat (modified from Jepsen, 1970). (m and n) *Pakicetus attocki*, one of the earliest whales from the early Middle Eocene of Pakistan (after Gingerich and Russell, 1981). (o) Occlusal view of right molar teeth (M_1–M_3) of a primitive raoellid *Khirtharia dayi* (after Thewissen et al., 2001)

Kumar, 1974; Kumar, 2000) and Kuldana and Domanda Formations of Pakistan. The following groups of tetrapods including squamates (Prasad and Bajpai, 2008), and diverse assemblages of mammals (Bajpai et al., 2005, 2006; Smith et al., 2007; Clyde et al., 2003; Gingerich et al., 1997, 1998, 1999, 2001; Thewissen et al., 1983, 2001) are known from the Early and Middle Eocene deposits of Indo-Pakistan subcontinent (Table 7.2; Fig. 7.9).

Table 7.2 Comparisons of Early and Middle Eocene tetrapod faunas of Indian subcontinent with Gondwana and Laurasia for palaeobiogeographic relationships

	Gondwana	Laurasia
AMPHIBIA		
Discoglossidae	+	
?Hyloidea or Ranoidea*	+	
REPTILIA		
Turtles		
Trionychidae		+
Trionyx		
Squamates		
Agamidae*		
Vastanagama susani	+	
Anguidae*		
Rajaurosaurus estesi	+	
Palaeophiidae*:		
Pterosphenus kutchensis	+	
Pterosphenus biswasi	+	
?Madtsoiidae	+	
or Boidae		
Colubroidea*		
Crocodiles		
Crocodylidae	+	+
Pristichampus		+
Crocodylus	+	
MAMMALIA		
Marsupials		
Didelphidae		+
Indodelphis luoi	+	
Jaegeria cambayensis	+	
Placentals		
Protoeutheria		
Plesiosoricidae*	+	
Pakilestes lathrious		+
Lipotyphla		
Perizalambdodon	+	
Seia shahi	+	
Condylartha		
Quettacyonidae*	+	+
Quettacyon parachai	+	
Machocyon abbasi	+	
Sororocyon usmanii	+	
Obashtakaia aeruginis	+	
Proboscidea		
Anthracobunidae*		
Nakusia shahrigensis	+	
Anthracobune pinfoldi	+	
Anthracobune aijensis	+	
Jozaria palustris	+	
Pilgrimella wardi	+	
Mesonychia		

Table 7.2 (continued)

	Gondwana	Laurasia
Mesonychidae		+
Tillodonta		
Estonychidae		+
Basalina basalensis		
Creodonta		
Hyaenodontidae		+
Paratritemnodon indicus	+	
Paratritemnodon jandewalensis	+	
Artocyonia		
Artocyonidae*		
Karakia longidens	+	
Perissodactyla		
Brontotheriidae		+
Eototanops dayi		+
Pakotitanops latidentatus		
Mulkrajamops moghliensis		
Isectolophidae	+	
Karagalax mamikhelensis	+	
Sastrilophus dehmi	+	
Chalicotheriidae		+
Helaletidae*	+	
Hyrachyus asiaticus		
Jhagirlophus chorgalensis		
Triplopus kalakotensis		
Depertellidae	+	
Teleohus daviesi		
Hyracodontidae		+
"Forstercooperia" jigniensis		
Equidae		+
Cambaytheriidae	+	
Cambaytherium thewissi		
Cambayherium minor		
Kalitherium marinus		
Indobune vastanensis		
Artiodactyla		
Dichobunidae*	+	
Chorlakkia hassani		
Dulcidon gandaensis		
Pakibune chorlakkiensis		
Raoellidae*	+	
Khirtharia dayi		
Khirtharia aurea		
Khirtharia inflata		
Haqueina haquei		
Indohyus indirae		
Indohyus major		
Kunmunella kalakotensis		

Table 7.2 (continued)

	Gondwana	Laurasia
Kunmunella transversa		
Metkatius kashmirensis		
Cetaceans		
Pakicetidae*	+	
Pakicetus attocki		
Himalayaectus subathuensis		
Gandakasia		
Ichthyolestes		
Rodentia		
Chapattimyidae		+
Chapattimus debruuijni		
Chapattimus wilsoni		
Chapattimus ibrahimshahi		
Saykanomys sondaari		
Sakanomys ijlsti		
Gumbatomys asifi		
Advenimus bohlini		
Ctenodactylidae		+
Paramyidae		+
Birbalomys woodi		
Metkamys blacki		
Yuomuidae		+
Advenimus bohlini		
Cf. *Petrokoslovia*		
Primates		+
Adapidae		+
Panobius afridi		
Cf. *Agerinia*		
Omomyidae*	+	
Kohatius coppensi		
Chiroptera		
Icaronycteridae*		+
Icaronycteris sigei		
Archaeonycteridae*	+	
Protonycteris gunnelli		
Archaeonycteris? storchi		
Hassianycterididae*	+	
Hassianycteris kumari		
Cambaya complexus		
Palaeochiropterygidae*		+
Microchiropteryx folieae		
Cimolesta		
Palaeorictydae		+
Anthraryctes vastanensis		
Climolestidae		+
Suratilester gingerichi		
Apatemyidae		+
Frugivastodon cristatus		

Table 7.2 (continued)

	Gondwana	Laurasia
Insectivores		
Family indet.	+	
Vasanta sania		
Cambay complexus		
Perizalamdodon punjabensis		

Faunal Index: *43% endemic or Indian origin; 55% Laurasian origin; 2% Gondwana origin.

7.4.2 Biogeography of the Eocene Tetrapod Faunas

The composition of the Eocene mammals from Indo-Pakistan differs greatly from that of Late Cretaceous period, suggesting that Indo-Pakistan was carrying its own peculiar fauna (Russell and Zhai, 1987) and finally converging to Asia, allowing faunal interchange. There is a dramatic shift from Gondwanan to Laurasian relationships in this early Eocene fauna. Out-of 42 family level taxa, only one family, ?Madtsoiidae, might have some Gondwana ties, 18 families may be endemic or ancestral to major clades (faunal index 43%), whereas 23 families are shared between India and Asia (faunal index 55%).

7.5 Discussion and Conclusion

New biogeographic synthesis suggests that the Late Cretaceous Indian tetrapod fauna is cosmopolitan with both Gondwanan and Laurasian elements by the formation of multiple temporary land bridges and is dominated by geodispersal biogeography. The Early Eocene tetrapods, in contrast, show the influence of both geodispersal and vicariance. Several endemic families of archaic mammals in Early Eocene may indicate partial isolation of India. Middle Eocene vertebrates show strong influence of geodispersal biogeography as India made the initial contact with Asia with development of northeast biotic corridor.

Acknowledgments We thank Ashok Sahni for setting the stage on Indian palaeobiogeography. We thank J. G. M. Thewissen and two anonymous reviewers for providing useful comments and insights. We thank Jeff Martz for arranging some illustrations. Texas Tech University and PALAEOMAP Project of University of Texas at Arlington supported this research.

References

Bajpai, S, Kapur, VV, Thewissen, JGM, Das, DP, Tiwari, BN (2006) New Early Eocene Cambaythere (Perissodactyla, Mammalia) from the Vastan Lignite Mine (Gujarat, India) and an evaluation of Cambaythere relationships. J Palaeontol Soc Ind, 51:101–110.

Bajpai, S, Kapur, VV, Thewissen, JGM, Tiwari, BN, Das, DP (2005) First fossil marsupials from India: Early Eocene *Indodelphis* n. gen. and *Jaegeria* n. gen from Vastan Lignite Mine, District Surat, Gujarat. J Palaeontol Soc Ind, 50:147–151.

Barron, EJ, Harrison, CGA (1980) An analysis of past plate motions in South Atlantic and Indian Oceans. In: Davies, P, Runcorn, SK (eds) Mechanisms of continental drift and plate tectonics. Academic Press, London, UK.

Briggs, JC (1989) The historic biogeography of India: isolation or contact? Syst Biol, 38:322–332.

Chatterjee, S, Rudra, DK (1996) K/T events in India: impact, volcanism and dinosaur extinction. Mem Queensland Mus, 39:489–532.

Chatterjee, S, Scotese, CR (1999) The breakup Gondwana and the evolution of the Indian plate. In: Sahni, A, Loyal, RS (eds) Gondwana assembly: new issues and perspectives. Ind Natl Sci Acad, New Delhi, India.

Chatterjee, S, Scotese, C (2007) Biogeography of the Mesozoic Lepidosaurs on the wandering Indian plate. In: Carvalho, IS, Cassab, RCT, Schwanke, C, Carvalho, MA, Fernandes, ACS, Rodrigues, MAC, Carvalho, MSS, Arai, M, Oliveira, MEQ (eds) Paleontologia: Cenários de Vida. Editoria Interciência, Rio de Janeiro, Brazil.

Clift, PD, Degnan, PJ, Hannigan, R, Blusztajn, J (2000) Sedimentary and geochemical evolution of the Dras forearc basin, Indus suture, Ladakh Himalaya, India. Geol Soc Am Bull, 112:450–466.

Clyde, WC, Khan, IH, Gingerich, PD (2003) Stratigraphic response and mammalian dispersal during initial India-Asia collision: evidence from the Ghazij Formation, Balochistan, Pakistan. Geology, 31:1097–1100.

Duncan, RA (1981) Hotspots in the southern oceans – an absolute frame of reference for motion of the Gondwana continents. Tectonophysics, 74:29–42.

Gaffney, ES, Chatterjee, S, Rudra, DK (2001) *Kurmademys*, a new side-necked turtle (Pelomedusoides: Bothremydidae) from the Late Cretaceous of India. Am Mus Novit, 3321:1–16.

Gaffney, ES, Sahni, A, Schleich, H, Singh, SD, Srivastava, R (2003) *Sankuchemys*, a new side-necked turtle (Pelomedusoides: Bothremydidae) from the Late Cretaceous of India. Am Mus Novit, 3405:1–10.

Gingerich, PD, Abbas, SG, Arif, M (1997) Early Eocene *Quettacyon parachai* (Condylartha) from the Ghazij Formation of Baluchistan (Pakistan): Oldest Cenozoic land mammal from South Asia. J Vertebr Paleontol, 17:629–637.

Gingerich, PD, Arif, M, Khan, III, Abbas, SG (1998) First early Eocene land mammals from the upper Ghazij Formation of Sor Range, Baluchistan. In: Ghaznavi, MI, Raja, S, Hasan, MT (eds) Siwaliks of South Asia. Proceedings of the 3rd GEOASS workshop, Geol Surv Pakistan, Islamabad, Pakistan.

Gingerich, PD, Arif, M, Khan, IH, Clyde, WC, Bloch, JI (1999) *Mahocyon abbasi*, a new early Eocene quettacyonid (Mammalia, Condylartha) from the middle Ghazij Formation of Mach and Deghari coal fields, Baluchistan (Pakistan). Contr Mus Paleontol Univ Michigan, 30: 233–250.

Gingerich, PD, Arif, M, Khan, IH, Haq, MU, Bloch, JI, Clyde, WC, Gunnell, GF (2001) Gandhera quarry, a unique mammalian faunal assemblage from the early Eocene of Pakistan. In: Gunnell, GF (ed) Eocene vertebrates: unusual occurrence and rarely sampled habitats. Plenum Press, New York, NY.

Gingerich, PD, Russell, DE (1981) *Pakicetus attocki*, a new archaeocete (Mammalia, Cetacea) from the early-middle Eocene of Kuldana Formation of Kohat (Pakistan). Contr Mus Paleontol Univ Michigan, 25:235–246.

Glennie, KW, Boeuf, MGA, Hughes-Clarkle, MW, Moody-Stuart, M, Pilaar, WFH, Reinhardt, BM (1973) Late Cretaceous nappes in the Oman Mountains and their geological significance. Am Assoc Pet Geol Bull, 57:5–27.

Hallam, A (1972) Continental drift and the fossil record. Sci Am, 227(11):55–66.

Huene, FV, Matley, CA (1933) Cretaceous Saurischia and Ornithischia of the central provinces of India. Geol Surv Ind Palaeontol ind, 21:1–74.

Jaegger, JJ, Courtillot, V, Tapponnier, P (1989) Palaeontological view of the ages of the Deccan Traps, the Cretaceous-Tertiary boundary, and the India-Asia collision. Geology, 17: 316–319.

Jain, SL (1986) New pelomedusid turtle (Pleurodira: Chelonia) remains from Lameta Formation (Maastrichtian) at Dongargaon, Central India, and review of pelomedusids from India. J Paleontol Soc Ind, 31:63–75.

Jain, SL, Bandyopadhyay, S (1997) New titanosaurid (Dinosauria: Sauropoda) from the Late Cretaceous of Central India. J Vertebr Paleontol, 17:114–136.

Jepsen, GL (1970) Bat origins and evolution. In: Wimsatt, WA (ed) Biology of bats. Academic Press, New York, NY.

Khajuria, CK, Prasad, GVR (1998) Taphonomy of a Late Cretaceous mammal-bearing microvertebrate assemblage from the Deccan intertrappean beds of Naskal, peninsular India. Palaeogeogr Palaeoclimatol Palaeoecol, 137:153–172.

Khosla, A, Sahni, A (2003) Biodiversity during the Deccan volcanic eruptive episode. J Asian Earth Sci, 21:895–908.

Krause, DW, Prasad, GVR, von Koenigswald, W, Sahni, A, Grine, F (1997) Cosmopolitan among Gondwana Late Cretaceous mammals. Nature, 390:504–507.

Kumar, K (2000) Correlation of continental Eocene vertebrate localities in the Indian subcontinent. Himal Geol, 21:63–85.

Lieberman, BS (2000) Paleobiogeography. Kluwer Academic, Plenum Publishers, New York, NY.

Lieberman, BS (2003) Paleobiogeography: the relevance of fossils to biogeography. Ann Rev Ecol Evol Syst, 34:51–69.

McKenna, MCC (1973) Sweepstakes, filters, corridors, Noah's Arks, and beached Viking funeral ships in paleogeography. In: Tarling, DH, Runcorn, SK (eds) Implications of continental drift to the earth sciences. Academic Press, London, UK.

Molnar, P, Tapponnier, P (1977) The collision between India and Eurasia. Sci Am, 236(4): 30–41.

Nelson, G, Platnik, N (1981) Systematics and biogeography. Columbia University Press, New York, NY.

Noble, GK (1930) The fossil frogs of the intertrappean beds of Bombay, India. Am Mus Novit, 401:1–13.

Patriat, P, Achache, J (1984) India-Eurasia collision chronology has implications for crustal shortening and driving mechanism of plate. Nature, 311:615–621.

Prasad, GVR, Bajpai, S (2008) Agamid lizards from the Early Eocene of western India: oldest Cenozoic lizards from South Asia. Palaeontol Electron, 11:1–19.

Prasad, GVR, Jaeger, JJ, Sahni, A, Gheerbrant, E, Khajuria, CK (1994) Eutherian mammals from the Upper Cretaceous (Maastrichtian) Intertrappean Beds of Naskal, Andhra Pradesh, India. J Vertebr Paleontol, 14:260–277.

Prasad, GVR, Rage, JC (1991) A discoglossid frog in the Late Cretaceous (Maastrichtian) of India. Further evidence for a terrestrial route between India and Laurasia in the latest Cretaceous. C R Acad Sci Paris, 313:272–278.

Prasad, GVR, Rage, JC (1995) Amphibians and squamates from the Maastrichtian of Naskal, India. Cret Res, 16:95–107.

Prasad, GVR, Sahni, A (1988) First Cretaceous mammal from India. Nature, 332:638–640.

Prasad, GVR, Verma, O, Sahni, A, Parmar, V, Khoshla, A (2007) A Cretaceous hoofed mammal from India. Science, 318:937.

Rage, JC (1988) Gondwana, Tethys, and terrestrial vertebrates during the Mesozoic and Cainozoic. In: Audley-Charles, MG, Hallam, A (eds) Gondwana and tethys, Geol Soc Lond Spec Publ, 37:235–272. Oxford University Press, Oxford.

Rage, JC, Bajpai, S, Thewissen, JGM, Tiwari, BN (2003) Early Eocene snakes from Kutch, western India, with a review of Paleophiidae. Geodiversitas, 25:695–716.

Rana, RS, Sati, KK (2000) Late Cretaceous-Paleocene crocodilians from the Deccan Trap-associated sedimentary sequences of peninsular India. J Palaeontol Soc Ind, 45:123–136.

Rana, RS, Wilson, GP (2003) New Late Cretaceous mammals from the Intertrappean beds of Rangpur, India and paleobiogeographic framework. Acta Palaeontol Polonica, 48: 331–348.

Robertson, AHF, Searle, MP (1990) The northern Oman Tethyan continental margin: stratigraphy, structure, concepts and controversies. In: Robertson, AHF, Searle, MP, Ries, AC (eds) The geology and tectonics of the Oman region, Geol Soc Lond Spec Publ, 49:251–284, London, UK.

Rose, KD (2006) The beginning of the age of mammals. Johns Hopkins University Press, Baltimore, MD.

Russell, DE, Zhai, RJ (1987) The Paleogene of Asia; mammals and stratigraphy. Mus natl d'Hist nat Sci Terre Mem, 52:1–488.

Sahni, A, Bajpai, S (1988) Cretaceous-Tertiary boundary events: the fossils vertebrate, palaeomagnetic and radiometric evidence from peninsular India. J Geol Soc Ind, 32:382–396.

Sahni, A, Kumar, K (1974) Palaeogene palaeobiogeography of the Indian subcontinent. Palaeogeogr Palaeoclimatol Palaeoecol, 15:209–226.

Sahni, A, Kumar, K, Hatenberger, JL, Jaeger, JJ, Rage, JC, Sudre, J, Vainey Liaud, M (1982) Micoverte'bre's noubeau des Traps du Deccan (Inde): mise en e'vidence d'une voie de communicaion terrestre probable entre la Laurasie at l' Inde a la limite Cre'tace' – Tertiaire. Bull Soc Geol France, 24:1093–1099.

Sahni, A, Rana, RS, Prasad, GVR (1987) New evidence for the paleobiogeographic intercontinental Gondwana relationships based on Late Cretaceous-earliest Paleocene coastal faunas from Peninsular India. Gondwana Six Geophys Monogr, 41:207–218.

Sampson, SD, Witmer, LM, Forster, CA, Krause, DW, O'Connor, PM, Dodson, P, Ravoavy, F (1998) Predatory dinosaur remains from Madagascar: Implications for the Cretaceous biogeography of Gondwana. Science, 280:1048–1051.

Sengor, AMC (1990) A new model for the Late Paleozoic-Mesozoic tectonic evolution of Iran and implications for Oman. The geology and tectonics of Oman region. Geol Soc Lond Spec Publ, 49:797–831.

Shackleton, RM, Ries, AC (1990) Tectonics of the Masirah fault zone and eastern Oman. In: Robertson, AHF, Searle, MP, Ries, AC (eds) The geology and tectonics of the Oman region, Geol Soc Lond Spec Publ, 49:715–724, London, UK.

Sharland, PR, Casey, DM, Davies, RB, Simmons, MD, Sutcliffe, OE (2004) Arabian plate sequence stratigraphy – revisions to SP2. GeoArabia, 9:100–214.

Simpson, GG (1947) Holarctic mammalian faunas and continental relationships during the Cenozoic. Geol Soc Am Bull, 58:613–687.

Smith, AB (1988) Late Paleozoic biogeography of East Asia and palaeontological constraints on plate tectonic reconstruction. Phil Trans R Soc Lond, 326A:189–227.

Smith, T, Rana, RS, Missiaen, P, Rose, KD, Sahni, A, Singh, H, Singh, L (2007) High bat (Chiroptera) diversity in the Early Eocene of India. Naturwissenschaften, 94: 791–800.

Stampfli, GM, Borel, GD (2002) A plate tectonic model for the Paleozoic and Mesozoic constrained by dynamic plate boundary and restored synthetic oceanic isochrones. Earth Planet Sci Lett, 196:17–33.

Thewissen, JGM, Russell, DE, Gingerich, PD, Hussain, TS (1983) A new dichobunid artidactyle (Mammalia) from the Eocene of northwest Pakistan. Proc Koninkliijke Nedererlandse Acad Wetenschappen, B86:153–180.

Thewissen, JGM, Williams, EM, Hussain, ST (2001) Eocene mammal faunas from northern Indo-Pakistan. J Vertebr Paleontol, 21:347–366.

Upchurch, P, Hunn, CA, Norman, DB (2002) An analysis of dinosaurian biogeography: evidence for the existence of vicariance and dispersal patterns caused by geological events. Proc R Soc Lond, B269:613–621.

Van der Voo, R (1993) Paleomagnetism of the Atlantic, Tethys and Iapetus Oceans. Cambridge University Press, Cambridge, MA.

Warburton, J, Burnhill, TJ, Graham, RH, Isaac, KP (1990) The evolution of the Oman Mountains
    foreland basin. In: Robertson, AHF, Searle, MP, Ries, AC (eds) The geology and tectonics of
    the Oman region, Geol Soc Lond Spec Publ, 49:419–428, London, UK.
Wilson, JA, Malkani, MS, Gingerich, PD (2001) New crocodyliform (Reptilia: Mesoeucrocodylia)
    from the upper Cretaceous Pab Formation of Vitakri, Balochistan (Pakistan). Contr Mus
    Paleontol Univ Michigan, 30:321–336.

Index

S. Bandyopadhyay (ed.), *New Aspects of Mesozoic Biodiversity,*
Lecture Notes in Earth Sciences 132, DOI 10.1007/978-3-642-10311-7,
© Springer-Verlag Berlin Heidelberg 2010